Design and Development of Arc Welding Robot

Dr.S. Mahadevan

Dr.P. Sevvel

Published by

BONFRING®
Intellectual Integrity

Design and Development of Arc Welding Robot

9 789386 638038 >

ISBN 978-93-86638-03-8

Authors

Dr.S.Mahadevan

Dr.P.Sevvel

Bonfring

309, 2nd Floor, 5th Street Extension, Gandhipuram,

Coimbatore-641 012.

Tamilnadu, India.

E-mail: info@bonfring.org

Website: www.bonfring.org

Phone: 0422 4213231

Dedicated to

My Wife Tamilselvi

who made my life meaningful

&

all my Family Members

Chapter	Contents	Page No

CHAPTER 1
INTRODUCTION

Industrial Robotic Welding so far the most popular application of robotics in worldwide. In fact, there is a huge number of products that require welding operations in their assembly processes. The car industry is probably the most important example, with the spot and MIG/ MAG welding operations in the car body workshops of the assembly lines. Never the less, there are an increasing number of smaller businesses, client oriented, manufacturing small series or unique products designed for each client.

With the development of modern industrial technology, the requirements for the welding quality and production efficiency of the great engineering projects are more and more high. Welding production automation can improve the production efficiency and the welding quality at the same time, but the existing welding robot technology needs very high processing path precision [1]. In the manufacture and production process of parts, it does not avoid the processing and assembly deviation of work piece, especially in the field of construction machinery [2, 3] and shipbuilding [4], low processing and assembly precision severely restricts application ability and range of the industry robot technology, affect the production efficiency [5].

Multilayer butt welding, used on big structures requiring very strong welds, and the welding process is complex and difficult to parameterize and control. Welding in most cases (i.e., MIG/MAG welding) imposes extremely high temperatures concentrated in small zones. Physically, that makes the material experience extremely high and localized thermal expansion and contraction cycles. This introduces changes in the material that may affect mechanical behavior along with plastic deformation. These changes must be well known in order to minimize the effects.

1.1. Gas Metal Arc Welding (GMAW)

Gas-metal-arc (GMA) fillet welding is one of the most commonly used welding processes in the industry [6]. MIG/MAG welding process, also known as gas metal arc welding process, uses the heat of the electric arc to melt the electrode wire and the metallic components to be welded. Figure 1.1 illustrates the welding principle. The fusion is carried out under the protection of a gas, or mixture of gases, in order to prevent pernicious contamination with atmospheric gases (oxygen, nitrogen, and hydrogen). The stability of the welding process is sensitive to basic welding parameters, especially current, voltage, welding speed, stick-out (length of wire out of the contact tube), shielding gas, and arc length.

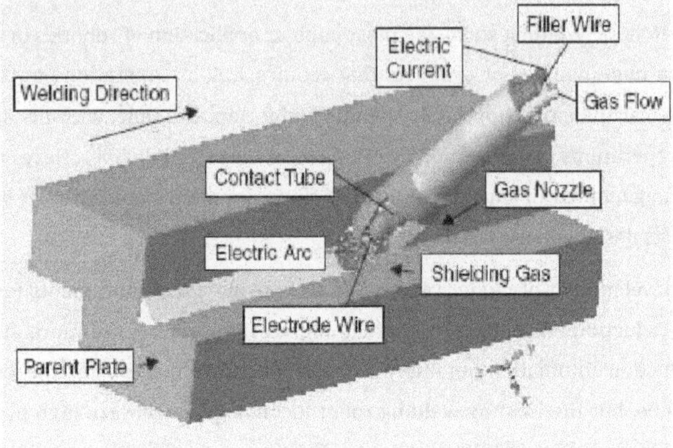

Fig. 1.1: The GMAW Process Principle

A small change in the distance between the welding torch and the component being welded may produce a considerable variation on current and voltage. Current, voltage, and shielding gas influence the transfer mode of melted filler wire to the component being welded, affecting the quality of the welds. If the electric arc is unstable defects like bad penetration profile, undercut, or excessive spatter, may occur.

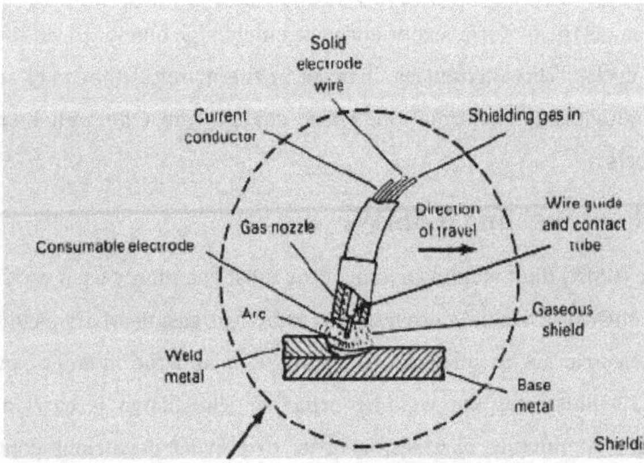

Fig. 1.2: Detail View of MIG/MAG

1.2. Welding Automation

Industrial robotic welding is so far the most popular application of robotics worldwide. This should lead to cheaper products since productivity and quality can be increased, and production costs and manpower can be decreased. The introduction of robotic is must wherever the manual seriously affects with hazardous environment.

Welding Parameter Calculations

Five (5) Factors of to decide the Arc Welding quality

1. Heat.
2. Electrode.
3. Electrode angle.
4. Arc length.
5. Speed of travel.

1.3. Welding Safety

The major hazards of concern during GMAW are the fumes and gases, which can harm health the high-voltage electricity, which can injure and kill the arc rays, which can injure eyes and burn skin and the noise which may be present that can damage hearing. The type and amount of fumes and gas present during welding depend on the electrode being used, the alloy being welded, and the presence of any coatings on the base metal.

Apart from safety awareness, the quality of weld can be achieved by proper fatigue free position and skill of the operator is so important. In this work the factors influencing for quality weld is worse affected due to improper position and hazardous environment. This is an overcome by using robotic welding.

CHAPTER 2

DESIGN OF ROBOT

2.1. Kinematics of Robot

The forward kinematics problem is to determine the position and orientation of the end-effectors, given the values for the joint variables of the robot. The joint variables are the angles between the links in the case of revolute or rotational joints, and the link extension in the case of prismatic or sliding joints.

2.1.1. Denavit Hartenberg Representation

While it is possible to carry out all of the analysis in this chapter using an arbitrary frame attached to each link, it is helpful to be systematic in the choice of these frames. A commonly used convention for selecting frames of reference in robotic applications is the Denavit-Hartenberg, or D-H convention [7]. In this convention, each homogeneous transformation Ai is represented as a product of four basic transformations

Ai = Transz, diTransx, aiRotx,_i Rotz,_i

2.1.2. Assignment of Coordinate Frames

For a given robot manipulator, one can always choose the frames 0 to n in such a way that the above two conditions are satisfied. In certain circumstances, this will require placing the origin oi of frame i in a location that may not be intuitively satisfying, but typically this will not be the case. In reading the material below, it is important to keep in mind that the choices of the various coordinate frames are not unique, even when constrained by the requirements above. Thus, it is possible that different engineers will derive differing, but equally correct, coordinate frame assignments for the links of the robot. It is very important to note, however, that the end result will be the same, regardless of the assignment of intermediate link frames (assuming that the coordinate frames for link n coincide). We will begin by deriving the general procedure. We will then discuss various common special cases where it is possible to further simplify the homogeneous transformation matrix. the above procedure based on the D-H convention in the following algorithm for deriving the forward kinematics for any manipulator.

Step l: Locate and label the joint axes z0, . . . ,zn−1.

Step 2: Establish the base frame. Set the origin anywhere on the z0-axis. The x0 and y0 axes are chosen conveniently to form a right-hand frame.

For i = 1, . . . , n – 1, perform Steps 3 to 5.

Step 3: Locate the origin oi where the common normal to zi and zi−1 intersects zi. If zi intersects zi−1 locate oi at this intersection. If zi and zi−1 are parallel, locate oi in any convenient position along zi.

Step 4: Establish xi along the common normal between zi−1 and zi through oi, or in the direction normal to the zi−1 – zi plane if zi−1 and zi intersect.

Step 5: Establish yi to complete a right-hand frame.

Step 6: Establish the end-effect or frame on xnynzn. Assuming the n-th joint is revolute, set zn = a along the direction zn−1. Establish the origin on conveniently along zn, preferably at the center of the gripper or at the tip of any tool that the manipulator may be carrying. Set yn = s in the direction of the gripper closure and set xn = n as s × a. If the tool is not a simple gripper set xn and yn conveniently to form a right- hand frame.

Step 7: Create a table of link parameters ai, di, αi, θi.

ai = distance along xi from oi to the intersection of the xi and zi−1 axes.

di = distance along zi−1 from oi−1 to the intersection of the xi and zi−1 axes. di is variable if joint i is prismatic.

αi = the angle between zi−1 and zi measured about xi (see Figure 2.1).

θi = the angle between xi−1 and xi measured about zi−1 (see Figure 2.1). θi is variable if joint i is revolute.

Step 8: Form the homogeneous transformation matrices Ai by substituting the above parameters into DH table.

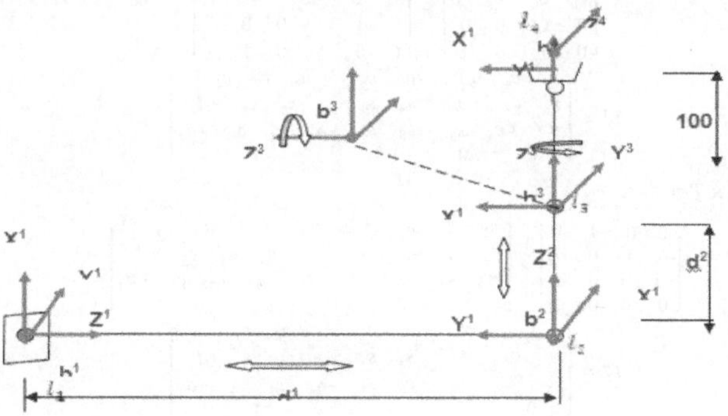

Fig. 2.1: Kinematics of Robot

Step 9: Form T0 n = A1 · · ·An. This then gives the position and orientation of the tool frame expressed in base coordinates.

Table 2.1: Denavat Hertenberg (DH) Parameters

Joints	Joint angle(θ)	Link dist(d)	Joint dist(a)	Link twist angle (α)	Home po- sition
1	0	q_1	0	$-\pi/2$	1800
2	$\pi/2$	q_2	0	$\pi/2$	120
3	q_3	0	0	$\pi/2$	0
4	q_4	0	a4	$\pi/2$	0

Where the four quantities $\theta_i, d_i, a_i, \alpha_I$ are parameters associated with link i and joint i. The four parameters $\theta_i, d_i, a_i, \alpha_I$ in (Table 2.1) are generally given the names link length, link twist, link offset, and joint angle, respectively. These names derive from specific aspects of the geometric relationship between two coordinate frames, as will become apparent below. Since the matrix Ai is a function of a single variable, it turns out that three of the above four quantities are constant for a given link, while the fourth parameter, θi for a revolute joint and di for a prismatic joint, is the joint variable.

$$T_{k-1}^{k} = \begin{bmatrix} \cos\theta_k & -\cos\alpha_k * \sin\theta_k & \sin\alpha_k * \sin\theta_k & a_k * \cos\theta_k \\ \sin\theta_k & \cos\alpha_k * \cos\theta_k & -\sin\alpha_k * \cos\theta_k & a_k * \cos\theta_k \\ 0 & \sin\alpha_k & \cos\alpha_k & d_k \\ 0 & 0 & 0 & 1 \end{bmatrix}$$

If K= 1, 2, 3 Then

$$T_0^2 = T_0^1 \times T_1^2 \begin{bmatrix} 1 & 0 & 0 & 0 \\ 0 & 0 & 1 & 0 \\ 0 & -1 & 0 & d_1 \\ 0 & 0 & 0 & 1 \end{bmatrix} \times \begin{bmatrix} 0 & 0 & 1 & 0 \\ 1 & 0 & 0 & 0 \\ 0 & 1 & 0 & d_2 \\ 0 & 0 & 0 & 1 \end{bmatrix} = \begin{bmatrix} 0 & 0 & 1 & 0 \\ 0 & 1 & 0 & d_2 \\ -1 & 0 & 0 & d_1 \\ 0 & 0 & 0 & 1 \end{bmatrix}$$

$$T_2^4 = T_2^3 \times T_3^4$$

$$\begin{bmatrix} c_3 & 0 & s_3 & 0 \\ s_3 & 0 & -c_3 & 0 \\ 0 & 1 & 0 & 0 \\ 0 & 0 & 0 & 1 \end{bmatrix} \times \begin{bmatrix} c_4 & 0 & s_4 & a_4 c_4 \\ s_4 & 0 & -c_4 & -a_4 s_4 \\ 0 & 1 & 0 & 0 \\ 0 & 0 & 0 & 1 \end{bmatrix}$$

$$T_3^4 = \begin{bmatrix} c_3 c_4 & s_3 & c_3 s_4 & a_4 c_3 c_4 \\ s_3 c_4 & -c_3 & s_3 s_4 & a_4 s_3 c_4 + d_2 \\ -c_3 c_4 & -s_3 & -c_3 s_4 & -a_4 c_3 c_4 + d_1 \\ 0 & 0 & 0 & 1 \end{bmatrix}$$

$$T_{base}^{tool}(T_0^4) = T_0^2 \times T_2^4$$

$$\begin{bmatrix} 0 & 0 & 1 & 0 \\ 0 & 1 & 0 & d_2 \\ -1 & 0 & 0 & d_1 \\ 0 & 0 & 0 & 1 \end{bmatrix} \times \begin{bmatrix} c_3 c_4 & s_3 & c_3 s_4 & a_4 c_3 c_4 \\ s_3 c_4 & -c_3 & s_3 s_4 & a_4 s_3 c_4 + d_2 \\ -c_3 c_4 & -s_3 & -c_3 s_4 & -a_4 c_3 c_4 + d_1 \\ 0 & 0 & 0 & 1 \end{bmatrix}$$

$$T_2^4 = \begin{bmatrix} s_4 & 0 & -c_4 & a_4 s_4 \\ s_3 c_4 & -c_3 & s_3 s_4 & a_4 s_3 c_4 + d_2 \\ -c_3 c_4 & -s_3 & -c_3 s_4 & -a_4 c_3 c_4 + d_1 \\ 0 & 0 & 0 & 1 \end{bmatrix}$$

Work envelope (Four axis)

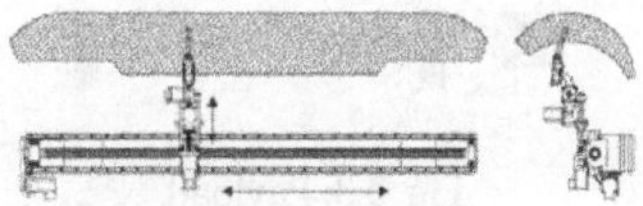

Fig. 2.2: Work Envelope

2.2. Design Approach

Selection of Ball screw and motor torque ensures the dynamics of robot. Here the selection of ball screw with relation to motor power to be evaluated with linear motion guide slides.

2.2.1. Selection Conditions

Payload	=	50kg
Stroke length(Ls)	=	2000mm
Maximum speed Vmax	=	9000mm/min(Rapid)
Acceleration time (t1)	=	0.2sec
Deceleration time (t3)	=	0.2sec
Positioning accuracy	=	±0.3mm/1000mm
Positioning Repeatability	=	±0.1mm
Minimum feed amount(s)	=	5mm/min
Driving AC servo motor		
Rated rotational speed	=	1000rpm
Reduction gear(A)	=	1
Guide surface resistance(f)	=	15 N (withoutload)

Lead (Ph)

$$Ph = \frac{V_{max}}{N_{max}} \text{ ....(1)}$$

Vmax : Maximum velocity (mm/min)
Nmax: Maximum speed of Ball Screw Shaft (rpm)

$\frac{9000}{1000}$ = lead 9mm and above

Brief Description

Pay load: The pay load has been calculated from the masses of following parts viz.,transfer block, cross slide, cross slide motors, welding torch etc. and its weight is around 10 Kg. but in account of possible increase in pay load in future and safe design selected it as 20 Kg. The feeder wire mounted quite opposite of torch assy and the load for main drive increases by 30 kg, hence total pay load for main drive is 50 kg.

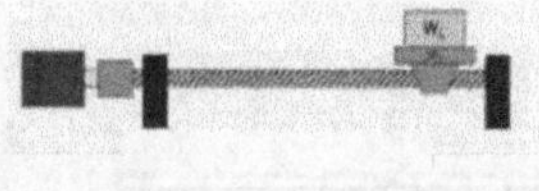

Fig. 2.3: Payload

Stroke length: The stroke length on X axis is 2000 mm and Y axis is 250 mm.

In order to meet that requirement, the ball screw length can be calculated as, Ls = Stroke +Nut length + margin + shaft end length

= 2000+64+40+120 = 2220mm, Similarly for Y axis Ls = 330mm

For rapid mode (teach-in), maximum linear speed chosen is 9000 mm/min, selected servo motor maximum rotational speed is 1000 rpm, hence minimum lead for the ball screw shaft should be 9 mm. these selection conforms the catalogue provided by the manufacturer. All the friction value at the places where relative motion is there are of rolling friction in nature, and while designing all these have been taken care, followed by the catalogues provided by the manufacturers.

A. Selection of Ball Screw Shaft

Ball screw drives the entire payload where the dynamic load factors to be analyzed and it should be withstand higher speed accordingly motor drives to be selected. Selection of ball screw is shown in Fig 2.4.

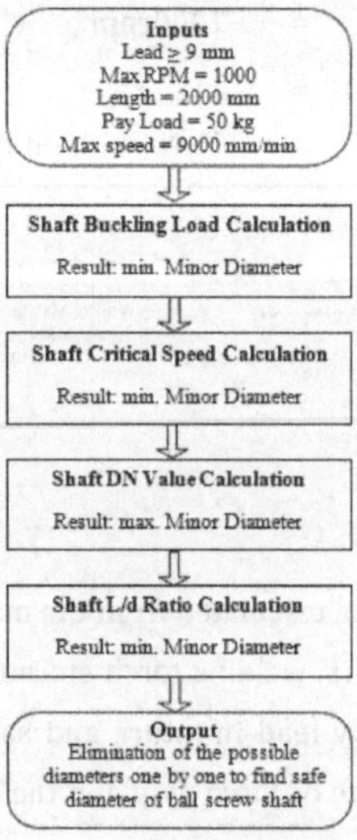

Fig. 2.4: Selection Process of Ball Screw

For safe design against shaft buckling load it is assumed that ball screw shaft is placed vertical and the pay load slides over it in vertical direction (forces will be maximum in this situation). One end of the ball screw shaft is assumed 'Fixed' and other end is 'Supported' (as per bearing supporting condition).

$$\text{Acceleration}(\alpha) = \frac{V_{max}}{t} = \frac{9000}{60*0.2*1000} = 0.75 m/\sec^2$$

Maximum axial loads in dynamic condition are found by

$Fa_1 = (m)\, g + (m)\, \alpha + f = 443\ N$ (acceleration) $= 40*9.81 + 40*.75 + 20 = 443$

$Fa_1 = (m)\, g - (m)\, \alpha - f = 343\ N$ (deceleration) $= 393 - 30 - 20 = 343$

A.1 Motion Profile

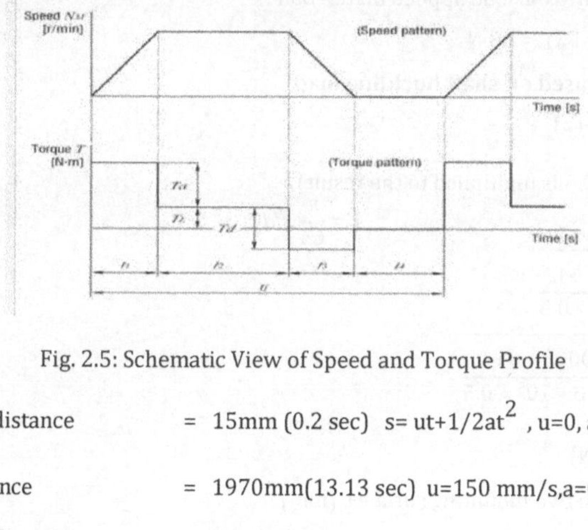

Fig. 2.5: Schematic View of Speed and Torque Profile

Acceleration distance = 15mm (0.2 sec) $s = ut + 1/2at^2$, u=0, a= 0.75 s =15mm

Uniform distance = 1970mm(13.13 sec) u=150 mm/s,a=0,$s = ut + 1/2at^2$

Deceleration distance = 15mm (0.2sec)

Dwelltime = 2sec

Maximum axial loads in dynamic condition are found by

1. During forward acceleration:

$Fa1 = \mu \bullet (m)\, g + f + (m) \bullet \alpha = 0.02(50)9.81 + 15 + (50)0.75 = 63\ N$

2. During forward uniform motion:

$Fa2 = \mu \bullet (m)\, g + f$

$= 0.02(50)9.81 + 15 = 25N$

3. During forward deceleration:

$Fa3 = \mu\bullet (m) g + f - (m)\bullet\alpha$

$= 0.02(50)9.81+15-(50)0.75 = -13N$

4. During backward acceleration:

$Fa4 = -\mu\bullet (m) g - f - (m)\bullet\alpha$

$= -0.02(50)9.81-15-(50)0.75 = -63$

5. During uniform backward motion:

$Fa5 = -\mu\bullet (m) g - f$

$= -0.02(50)9.81-15 = -25N$

6. During backward deceleration:

$Fa6 = -\mu\bullet (m) g - f + (m)\bullet\alpha$

$= -0.02(50)9.81 - 15 + (50)0.75 = 13$

Thus, the maximum axial load applied on the Ball

Screw is: Fa max = Fa1 = 63 N

Step 1: Selection based on shaft buckling load

$P_1 = \frac{\eta_1 * \pi^2 * E * I}{l_a^2} \times 0.5$ --------(2)

(A safety factor of 0.5 is multiplied to the result)

$$I = \frac{\pi}{64} \times d_1^4$$

$$d1 \geq \sqrt[4]{\frac{P_1 * l_a{}^2 * 64}{n_1 * \pi^3 * E * 0.5}}$$

$$d1 \geq \sqrt[4]{\frac{443 * 2000^2 * 64}{2 * \pi^3 * 2.06 * 10^5 * 0.5}}$$

d1 ≥ 11.41 mm

P1: Buckling load (N)

l_a: Distance between two mounting surfaces (mm)

E: Young's modulus($2.06*10^5 N/mm^2$)

I: moment of inertia of the shaft (mm^4)

d1 = Screw thread minor diameter

η1 = Factor according to the mounting method

Fixed–free	η1 =0.25
Fixed–supported	η1 =2
Fixed–fixed	η1 = 4

Step 2: Selection based on dangerous speed of the screw shaft (N1)

$N_1 = \frac{60 * \lambda_1^2}{2\pi * l_b^2} \times \sqrt{\frac{E * 10^3 * I}{\gamma * A}} \times 0.8$---------------(3)

(A safety factor of 0.8 is multiplied to the result)

$$d_1 = \frac{2\pi * l_b^2 * N}{\lambda_1^2 * 60 * 0.8} * \left(\frac{16 * \gamma}{E * 10^3}\right)$$

$$d_1 = \frac{2\pi * 2000^2 * 1000}{3.927^2 * 60 * 0.8} * \left(\frac{16 * 7.85 * 10^6}{2.06 * 10^5 * 10^3}\right)$$

$= 26.51$

$d_1 \geq 26.51$ mm

N1: Permissible rotational speed determined by the dangerous speed (RPM)

l_b: Distance between two mounting surfaces (mm)

E: Young's modulus($2.06 * 10^5$ N/mm^2)

I: Minimum geometrical moment of inertia of the shaft (mm^4)

d_1: Screw thread minor diameter

γ: Density (specific gravity)($7.85 * 10^{-6}$ kg/mm^3)

A: Screw shaft cross-sectional area(mm^2)

λ1: Factor according to the mounting method

Fixed–free λ1=1.875

Supported – supported λ1 = 3.142

Fixed–supported λ1=3.927

Fixed–fixed λ1 =4.73

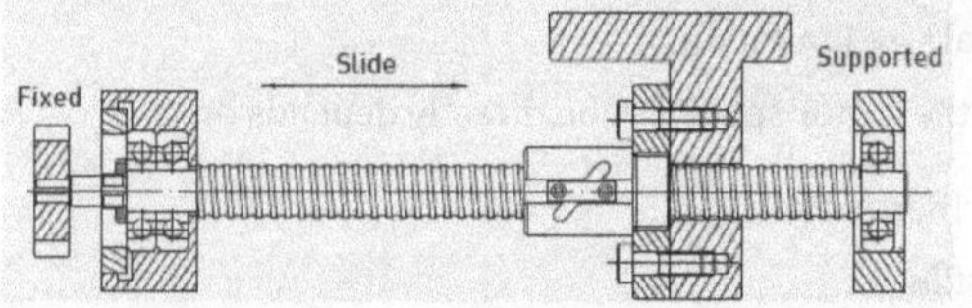

Fig. 2.6: Layout of Fixed – Supported Configuration

Step 3: Selection based on DN Value

DN ≤ 50000

$$d_1 \leq \frac{50000}{1000}$$

$d_1 \leq 50$ mm

d_1 should be 25,28,32,36,40

N2: Permissible rotational speed detemined by the DN value (rpm)

d_1: Screw thread minor diameter

Step 4 : Selection based on L/d Ratio

Normally ls/d (screw length / shaft diameter) ≤ 70 is recommended

$$d = \geq \frac{Ls}{70} \geq \frac{2000}{70}$$

d≥ 28.57 mm

Screw shaft diameter = 32, 36, 40 mm and lead 10mm

Next 28 mm as per calculation 32 mm outer dia and lead 10 mm selected for main drive of ball screw.

B. Selection of AC Servomotor

Sizing of motor involves the following factors

1. Inertia
2. Motion profile
3. Speed
4. Acceleration
5. Torque

In addition to that other factors also be considered while do the selection of the motor, they are

a. Cost
b. Resolution
c. Environment
d. Power requirement
e. Physical size limitations

The outcome of the motor specification directly depends on

1. Moment of Inertia ratio
2. Motion profile
3. Max torque @speed
4. RMS torque @speed

The selection of driving motor required to rotate the ball screw, normally take into account of the rotational speed, rotational torque and minimum feed amount.

Moment of Inertia Ratio(MIR)

$$MIR = \frac{Load\ Inertia}{Motor\ Inertia}$$

Load inertia contributes the ball screw and the work load

a) Inertia of Ball Screw(JB)

$$J_B = \frac{1}{2g} \times W_s \times \frac{D_N^2}{2} \times \frac{N_1^2}{N_2}$$

$$J_B = \frac{1}{2 * 9800} \times 14 \times \frac{32^2}{2} \times 1^2$$

$$J_B = 0.183\ kgf.mm - sec^2$$

JB : Inertia of ball screw kgf.mm-sec^2)

Ws : Ball screw weight(kgf)

DN : Ball screw nominaldia(mm)

b) Inertia of Workload(JL)

$$J_L = \frac{w}{g} \times \frac{ph^2}{2\pi} \times \frac{N_1}{N_2}^2$$

$$J_L = \frac{50}{9800} \times \frac{10^2}{2\pi} \times 1^2$$

$J_L = 0.013\ kgf.mm - sec^2$

Total load Inertia (JL Total)

$\boldsymbol{J_L Total = J_B + J_L}$

= 0.183 + 0.013

$J_L Total = 0.196\ kgf.mm - sec^2$

Motor Inertia

$$J_M = \frac{1}{2g} \times \pi * \rho_d * R^4 * L$$

$$J_M = \frac{1}{2 * 9800} \times \pi * 7.8 * 10^{-6} * 25^4 * 80$$

$J_M = 0.04\ kgf.mm - sec^2$

JM : Motor inertia (kgf-mm-sec^2)

ρd : Shaft specific weight (7.81x10^{-6}kgf-mm)

R : Disc radius(mm)

L : Disc length(mm)

g : gravity coefficient (9800mm/sec^2)

Moment of Inertia Ratio (MIR)

$$MIR = \frac{0.196}{0.04} = 5$$

Inertia Ratio Guidelines

5:1 typical

2:1 or less for higher performance

10:1 or higher when performance not critical Due to MIR is 5 no need to add a gearbox

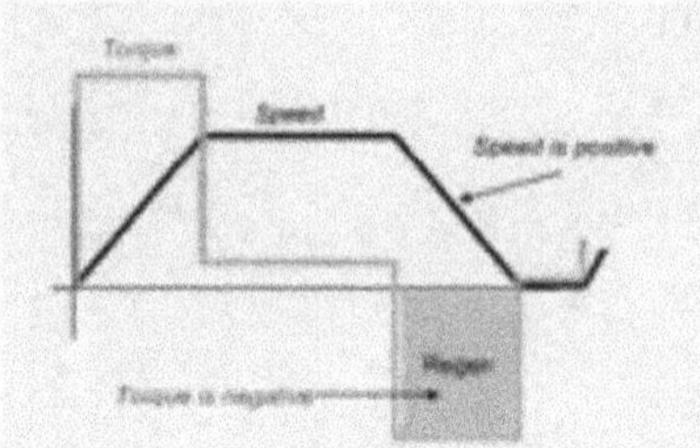

Fig. 2.7: Motion Profile

c) Torque Components

Friction and Acceleration

i. Friction

- Always present Baseline

ii. Acceleration

- Present while accelerating or decelerating
- Accelerating add to friction torque
- Decelerating substrates from friction torque

i. Friction Torque

a. Frictional Torque in Normal Drive Condition

$$T_a = \frac{F_{a1} * ph}{2\pi * \eta} \; N - mm$$

$$T_a = \frac{63 * 10}{2\pi * 0.8} \; N - mm$$

$$T_a = 125 \; N - mm = 0.125 \; N - m$$

Ta : Frictional torque due to an external load (N-mm)

Fa : Applied axial load (N)

Ph : Ball Screw lead (mm)

η : Ball Screw efficiency (0.8 to 0.9)

$$T_d = \frac{k_p * P * ph}{2\pi}$$

$$P = \frac{F_{a1}}{2.8} = \frac{63}{2.8} = 23N$$

$$T_d = \frac{0.2 * 23 * 10}{2\pi} = 7.3N - mm = 0.0073N - m$$

$$T_m = (T_a + T_b + T_d) \times \frac{D_1}{D_2}$$

$$T_m = (125 + 10 + 7.3) \times 1 = 142.3N - mm = 0.142N - m$$

Td : Preload drag torque (N-mm)

P: Preload (N)

Ph : Ball Screw lead (mm)

η : Ball Screw efficiency (0.8 to 0.9)

Kp : Preload torque coefficient

$$K_p = \frac{1}{\eta_1} - \eta_2 (is\ between\ 0.1 - 0.3)$$

ii. **Acceleration Torque(Ta)**

$$T_a^{'} = J \times \alpha$$

$$J_{Total} = J_L + J_M$$

$$J_{Total} = 0.196 + 0.04 = 0.236\ kgf.mm - sec^2$$

$$\alpha = \frac{2 * \pi * N}{60 * t} rad/sec^2$$

$$\alpha = \frac{2 * \pi * 1000}{60 * 0.2} rad/sec^2 = 524$$

$$T_a^{'} = 0.236 \times 524$$

$$T_a^{'} = 124\ kgf - mm = 1.24N - mm = 1240N - mm$$

TM : Motor drive torque (N-mm)

Tb : Friction torque of supporting torque (N-mm)

D1 : Diameter of driver pulley(mm)

D2 : Diameter of driven pulley(mm)

T'a : Motor drive torque at acceleration(kgf-mm)

J : System inertia(kgf-mm-sec^2)

α : Angular acceleration (rad/sec^2)

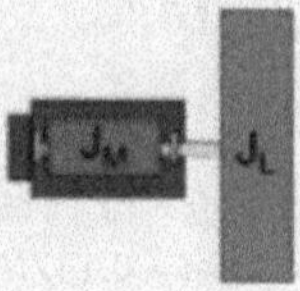

Fig. 2.8: Moment of Inertia

Total Motor Torque

$T_{Ma} = T_M + T_a^{'}$

$T_{Ma} = 142 + 1240 = 1382\, N - mm = 1.4N - m$

TMa : Total operating torque (kgf-mm)

Drive Power

$T_{pmax} = 2 \times 1382 = 2764N - mm2.8N - m(safety\, factor\, 2)$

$$P_d = \frac{2 * \pi * N * T_{pmax}}{60} = \frac{2 * \pi * 2.8 * 2000}{60} = 586W = 0.78hp$$

Selection of Motor

Thus the AC servo motor with following specification can be chosen

Rated torque: 1382N- mm = 1.4N-m

Rated rotational speed: 2000 rpm

Maximum torque: 2764N- mm = 2.8 N-m

The rotational motor torque of main shaft is calculated 2.8 N-m and drive power 0.8 hp and above, selected as 780W & rated torque 2.4N-m with rated rotational speed 3000rpm

Motion Profile

Motion profile the details of cycle time

Motion	Applied axial load Fa (N)	Travel distance Ln(mm)
No:1 During acceleration	63	15
No.2 During Forward uniform motion	25	1970
No.3 During Forward deceleration	-13	15
No.4 During backward acceleration	-63	15
No.5 During uniform backward motion	-25	1970
No.6 During backward deceleration	13	15

Maximum torque: 2764N- mm = 2.8N-m

Root Mean Square Torque @Speed

Effective Torque values can be calculated as RMS torque

Tma = TM+Ta = 142+1240 =1382N

t1 = 0.2sec

Tt = TM = 142N

t2 = 0.2sec

Ta = Ta-TM = 1098N

t3 = 0.2sec

Ts =0

t4 = 2.6 sec (dwelltime)

$$T_{rms} = \sqrt{\frac{T_{ma}^2 * t_1 + T_t^2 * t_2 + T_a^2 * t_3 + T_s^2 * t_4}{t_1 + t_2 + t_3 + t_4}}$$

$$T_{rms} = \sqrt{\frac{249^2 * 0.2 + 210^2 * 4.3 + 171^2 * 0.2 + 0}{0.2 + 4.3 + 0.2 + 1.6}} = 124N - mm = 0.124N - m$$

Similarly the cross slide ball screw and motor sizing is carried out as below.

For safe design against shaft buckling load it is assumed that ball screw shaft is placed vertical and the pay load slides over it in vertical direction (forces will be maximum in this situation). One end of the ball screw shaft is assumed 'Fixed' and other end is 'Free' (as per bearing supporting condition).

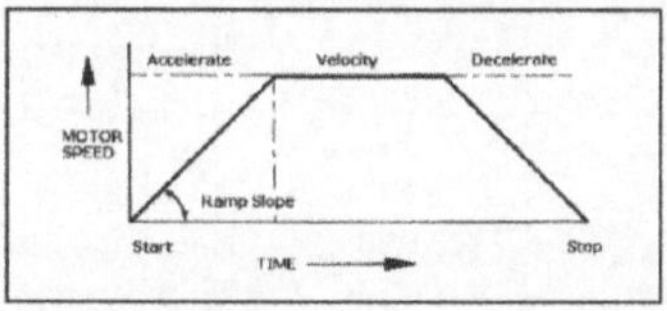

$$Acceleration(\alpha) = \frac{V_{max}}{t} = \frac{50}{60 * 0.2 * 1000} = 0.04m/sec^2$$

Maximum axial loads in dynamic condition are found by

Accl distance =42mm (0.2 sec)

Const distance =36mm (4.32 sec)

Decl distance = 42mm (0.2 sec)

Fa1 = (m) g+(m)α +f (acceleration) = 20*9.81+20*0.75+20 =231.2N

Fa1 = (m) g- (m)α -f (deceleration) = 20*9.81- 20*0.75 - 20 = 161.2N

f = LM Guide Rail surface resistance without load (N) = 20 N

Step 1: Selection based on shaft buckling load

$$P_1 = \frac{\eta_1 * \pi^2 * E * I}{l_a^2} \times 0.5$$

(A safety factor of 0.5 is multiplied to the result)

$$I = \frac{\pi}{64} \times d_1^4$$

$$d1 \geq \sqrt[4]{\frac{P_1 * l_a{}^2 * 64}{n_1 * \pi^3 * E * 0.5}}$$

$$d1 \geq \sqrt[4]{\frac{443 * 2000^2 * 64}{2 * \pi^3 * 2.06 * 10^5 * 0.5}}$$

$d1 \geq 11.41$ mm
P1: Buckling load (N)

l_a: Distance between two mounting surfaces (mm)

E: Young's modulus($2.06*10^5 N/mm^2$)

I: moment of inertia of the shaft (mm^4)

d1 = Screw thread minor diameter

η_1= Factor according to the mounting method

Fixed–free $\eta_1=0.25$

Fixed–supported $\eta_1=2$

Fixed–fixed $\eta_1=4$

Step 2: Selection based on dangerous speed of the screw shaft (N1)

$$N_1 = \frac{60 * \lambda_1^2}{2\pi * l_b^2} \times \sqrt{\frac{E * 10^3 * I}{\gamma * A}} \times 0.8$$

(A safety factor of 0.8 is multiplied to the result)

$$d_1 = \frac{2\pi * l_b^2 * N}{\lambda_1^2 * 60 * 0.8} \times \sqrt{\frac{16 * \gamma}{E * 10^3}}$$

$$d_1 = \frac{2\pi * 120^2 * 1000}{1.875^2 * 60 * 0.8} \times \sqrt{\frac{16 * 7.85 * 10^{-6}}{2.06 * 10^5 * 10^3}} = 6.5$$

$d_1 \geq 6.5mm$

N1: Permissible rotational speed determined by the dangerous speed (RPM)

l_b: Distance between two mounting surfaces(mm)

E: Young's modulus($2.06*10^5$ N/mm^2)

I: Minimum geometrical moment of inertia of the shaft (mm^4)

d1: Screw thread minor diameter

γ: Density (specific gravity)($7.85* 10^{-6} kg/mm^3$)

A: Screw shaft cross-sectional area(mm^2)

$\lambda1$: Factor according to the mounting method

Fixed–free $\lambda1=1.875$

Supported – supported $\lambda1 = 3.142$

Fixed – supported $\lambda1 = 3.927$

Fixed–fixed $\lambda1 =4.73$

Step 3: Selection based on DN value

DN≤50000

$$d_1 \le \frac{50000}{1000} = 50\ mm$$

d_1 should be 8, 10, 12

N2: Permissible rotational speed determined by the DN value (rpm)

d1: Screw thread minor diameter

Step 4 : Selection based on L/d ratio

Normally ls/d (screw length / shaft diameter) ≤ 70 is recommended

$$d \ge \frac{L_s}{70} \ge \frac{120}{70} \ge 2mm$$

Screw shaft diameter = **16** mm and lead **5** mm

C. Selection of AC servomotor

For Y axis sizing of motor involves the following factors

1. Moment of Inertia ratio
2. Motion profile
3. Max torque @speed
4. RMS torque @speed

The selection of driving motor required to rotate the ball screw, normally take into account of the rotational speed, rotational torque and minimum feed amount.

Moment of Inertia Ratio(MIR)

$$MIR = \frac{Load\ Inertia}{Motor\ Inertia}$$

Load inertia contributes the ball screw and the work load

a) Inertia of ball screw(JB)

$$J_B = \frac{1}{2g} \times W_s \times \frac{D_N}{2}^2 \times \frac{N_1}{N_2}^2$$

$$J_B = \frac{1}{2 * 9800} \times 0.5 \times \frac{16^2}{2} \times 1^2$$

$$J_B = 0.00163\ kgf.mm - sec^2$$

JB : Inertia of ball screw kgf.mm-sec^2)

W_S : Ball screw weight(kgf)

DN : Ball screw nominaldia(mm)

b) Inertia of Workload(JL)

$$J_L = \frac{w}{g} \times \frac{ph^2}{2\pi} \times \frac{N_1}{N_2}^2$$

$$J_L = \frac{25}{9800} \times \frac{5}{2\pi}^2 \times 1^2$$

$J_L = 0.00162\, kgf.mm - sec^2$

Total load Inertia (JL Total)

$\boldsymbol{J_L Total = J_B + J_L}$

= 0.00163 + 0.00162

$J_L Total = 0.00325\, kgf.mm - sec^2$

Motor Inertia

$$J_M = \frac{1}{2g} \times \quad \pi * \rho_d * R^4 * L$$

$$J_M = \frac{1}{2 * 9800} \times \quad \pi * 7.8 * 10^{-6} * 8^4 * 80$$

$J_M = 0.0004\, kgf.mm - sec^2$

JM : Motor inertia (kgf-mm-sec^2)

ρd : Shaft specific weight (7.81x10^{-6}kgf-mm)

R : Disc radius(mm)

L : Disc length(mm)

g : gravity coefficient (9800mm/sec^2)

Moment of Inertia Ratio (MIR)

$$MIR = \frac{Load\ Inertia}{Motor\ Inertia}$$

$$MIR = \frac{0.00325}{0.0004} = 8.125$$

Inertia Ratio Guidelines

5:1 typical

2:1 or less for higher performance

10:1 or higher when performance not critical

Due to MIR is 8 no need to add a **gearbox**

c) Torque Components

Friction and Acceleration

iii. Friction

- Always present Baseline

iv. Acceleration

- Present while accelerating or decelerating
- Accelerating add to friction torque
- Decelerating substrates from friction torque

i. Friction Torque

a. Frictional torque in normal drive condition

$$T_a = \frac{F_{a1} * ph}{2\pi * \eta} \; N - mm$$

$$T_a = \frac{27 * 5}{2\pi * 0.8} \; N - mm$$

$$T_a = 26.85 \; N - mm = 0.027 \; N - m$$

Ta : Frictional torque due to an external load (N-mm)

Fa : Applied axial load (N)

Ph : Ball Screw lead (mm)

η : Ball Screw efficiency (0.8 to 0.9)

$$T_d = \frac{k_p * P * ph}{2\pi}$$

$$P = \frac{F_{a1}}{2.8} = \frac{27}{2.8} = 9.64N$$

$$T_d = \frac{0.2 * 10 * 5}{2\pi} = 1.59N - mm = 0.0016N - m$$

$$T_m = (T_a + T_b + T_d) \times \frac{D_1}{D_2}$$

$$T_m = (27 + 10 + 1.6) \times \; 1 = 38.6N - mm = 0.039N - m$$

Td : Preload drag torque (N-mm)

P: Preload (N)

Ph : Ball Screw lead (mm)

η : Ball Screw efficiency (0.8 to 0.9)

Kp : Preload torque coefficient

$$K_p = \frac{1}{\eta_1} - \quad \eta_2 (is\ between\ 0.1 - 0.3)$$

iii. Acceleration Torque (Ta)

$T'_a = J \times \alpha$

$J_{Total} = J_L + J_M$

$J_{Total} = 0.00325196 + 0.0004 = 0.004\ kgf.mm - sec^2$

$$\alpha = \frac{2 * \pi * N}{60 * t} rad/sec^2$$

$$\alpha = \frac{2 * \pi * 1000}{60 * 0.2} rad/sec^2 = 524$$

$T'_a = 0.004 \times 524$

$T'_a = 2.1\ kgf - mm = 0.0021N - m = 210N - mm$

TM : Motor drive torque (N-mm)

Tb : Friction torque of supporting torque (N-mm)

D1 : Diameter of driver pulley(mm)

D2 : Diameter of driven pulley(mm)

T'a : Motor drive torque at acceleration(kgf-mm)

J : System inertia(kgf-mm-sec^2)

α : Angular acceleration (rad/sec^2)

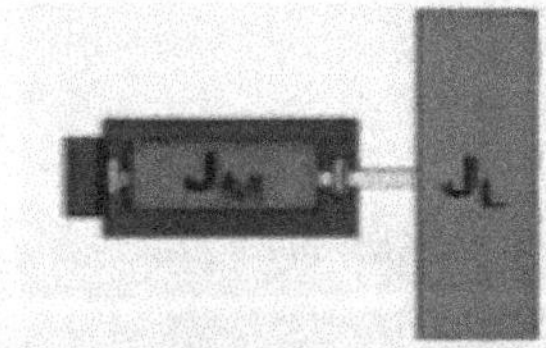

Moment of Inertia

Total Motor Torque

$T_{Ma} = T_M + T'_a$

$T_{Ma} = 39 + 210 = 249\ N - mm = 0.25N - m$

TMa : Total operating torque (kgf-mm)

Drive Power

$T_{pmax} = 2 \times 250 = 500N - mm2.8N - m(safety\ factor\ 2)$

$$P_d = \frac{2 * \pi * N * T_{pmax}}{60} = \frac{2 * \pi * 0.5 * 3000}{60} = 157W = 0.2hp$$

Selection of Motor

Thus the AC servo motor with following specification can be chosen

Rated torque: 250N- mm = 0.2N-m

Rated rotational speed: 3000 rpm

Maximum torque: 500N- mm = 0.5 N-m

The rotational Y axis motor torque of is calculated as 0.5 N-m and drive power 0.07 hp and above, selected as 400W & rated torque 2.4N-m with rated rotational speed 3000rpm

Motion Profile

Motion profile the details of cycle time

Maximum axial loads in dynamic condition are found by

1. During forward acceleration:

   $Fa1 = \mu \bullet (m)\, g + f + (m) \bullet \alpha$

   $= 0.02(25)9.81+20+(25)0.04 = 26$ N

2. During forward uniform motion:

   $Fa2 = \mu \bullet (m)\, g + f$

   $= 0.02(25)9.81+20 = 25$N

3. During forward deceleration:

   $Fa3 = \mu \bullet (m)\, g + f - (m) \bullet \alpha$

   $= 0.02(25)9.81+20-(25)0.04 = 24$N

4. During backward acceleration:

   $Fa4 = -\mu \bullet (m)\, g - f - (m) \bullet \alpha$

   $= -0.02(25)9.81-20-(25)0.04$

   $= -26$

5. During uniform backward motion:

   $Fa5 = -\mu \bullet (m)\, g - f$

   $= -0.02(25)9.81-20$

   $= -25$N

6. During backward deceleration:

   $Fa6 = -\mu \bullet (m)\, g - f + (m) \bullet \alpha$

   $= -0.02(25)9.81 - 20 + (25)0.04$

   $= -24$

Thus, the maximum axial load applied on the Ball Screw is:

$Fa\ max = Fa1 = 63$ N

Motion	Applied axial load Fa (N)	Travel distance Ln(mm)
No:1 During acceleration	26	42
No.2 During Forward uniform motion	25	36
No.3 During Forward deceleration	24	42
No.4 During backward acceleration	-26	42
No.5 During uniform backward motion	-25	36
No.6 During backward deceleration	-24	42

Maximum torque: 2764N- mm = 2.8N-m

Root Mean Square Torque @Speed

Effective Torque values can be calculated as RMS torque

Tma = TM+Ta = 39+210 =249N

t1 = 0.2sec

Tt = TM = 210N

t2 = 4.3sec

Ta = Ta-TM = 171N

t3 = 0.2sec

Ts =0

T4 = 1.6 sec (dwelltime)

$$T_{rms} = \sqrt{\frac{T_{ma}^2 * t_1 + T_t^2 * t_2 + T_a^2 * t_3 + T_s^2 * t_4}{t_1 + t_2 + t_3 + t_4}}$$

$$T_{rms} = \sqrt{\frac{249^2 * 0.2 + 210^2 * 4.3 + 171^2 * 0.2 + 0}{0.2 + 4.3 + 0.2 + 1.6}} = 124N - mm = 0.124N - m$$

2.3. Welding Installation and Procedures

Fig. 2.9: Welding Installation

Welding procedures emphasize control of parameters such as electrode speed (or current), voltage, welding speed, contact tube-to-base metal distance, as well as current pulse parameters for out-of-position welding. It is therefore easy to overlook the fact that the process is simply a source of heat and mass inputs to the weldment. Melting of the base metal, dilution of the filler metal, solidification of the weld bead, micro structural development in the weld bead and HAZ, and thermo mechanical distortion and residual stresses all follow from the heat and mass inputs. The conventional parameters identified above arc variables that control the heat and mass inputs

- Fix the robot at extremely close to the weld pool with maximum retract position of crossrail.
- Initial alignment to be done with joystick to ensure parallelism w.r.t weld pool.
- Teach-in Mode: Welding path to be programmed in teach-in mode[1]
- Welding mode: The advanced axis controller will manage the required path which ensures the constant feed and standoff distances

CHAPTER 3

ROBOT INSTALLATION AND PROCEDURES

3.1. Control System

The axis controller connects the each servo motor and it is controls the individual axis based on the reference command which is shown as block diagram in Fig. 3.1. This control system enables the closed loop system. Which is more helpful for identification of position and interrupts of welding can be done [8].

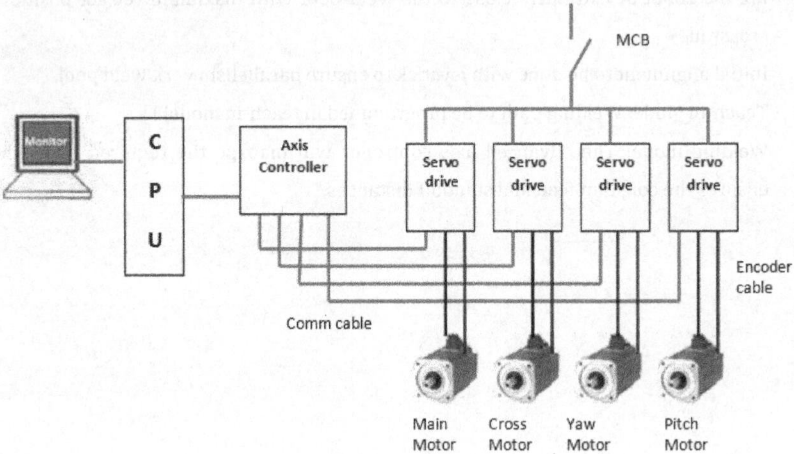

Fig. 3.1: Control System

This control system ensures the flexible program and direct access of cad drawings.

- Each servo motor has its own controller, providing real time closed-loop control.
- All the three controllers will be connected to Master Axis Controller via RS485 line.
- Master Axis Controller will be connected to PC/ Laptop using USB cable.
- A custom made software with GUI will be provided by the Axis controller manufacturer, for controlling all the three motors.
- The software will have Teach-in mode and Welding mode.
- Master Axis controller will shut down the system, in case the communication with PC breaks.

3.2. Motion Controller Architecture

A motion controller acts as the brain of the motion control system and calculates each commanded move trajectory. Because this task is vital, it requires a dedicated resource that offers a high level of determinism. The motion controller uses the trajectories it calculates to

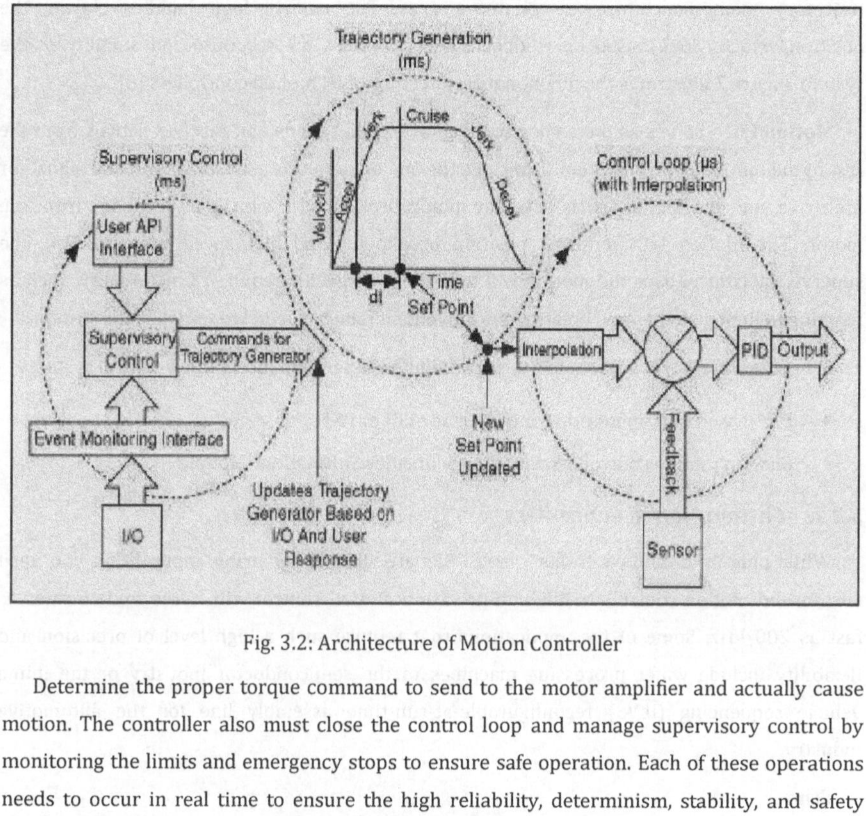

Fig. 3.2: Architecture of Motion Controller

Determine the proper torque command to send to the motor amplifier and actually cause motion. The controller also must close the control loop and manage supervisory control by monitoring the limits and emergency stops to ensure safe operation. Each of these operations needs to occur in real time to ensure the high reliability, determinism, stability, and safety necessary to create a working motion control system.

The following list describes the different tasks of a motion controller.

Supervisory control – Provides the command sequencing and coordination required to carry out specified operations such as the following:

- System initialization, which includes homing to a zero position.
- Event handling, which includes electronic gearing, triggering outputs based on position, and updating profiles based on user-defined events.

- Fault detection, which includes stopping moves on a limit switch encounter, safe system reaction to emergency stop or drive faults, watchdog, and so on.

Trajectory generator – Provides path planning based on the profile specified by the user.

Control loop – Performs fast, closed-loop control with simultaneous position, velocity, and trajectory maintenance on one or more axes. The control loop handles closing the position/velocity loop based on feedback, and it defines the response and stability of the system. Figure 7 illustrates the functional architecture of NI motion controllers [8].

Motion I/O - Serves as the analog and digital I/O that sends and receives signals from the rest of the motion control system. Typically, the analog output is used as a command signal for the drive, and the digital I/O is used for quadrature encoder signals as feedback from the motor. The motion I/O performs position breakpoint and high-speed capture. Also, the supervisory control uses the motion I/O to achieve a specific required functionality, such as reacting to limit switches and creating the movement modes needed to initialize the system.

Schneider offers two approaches to implement high-performance motion control systems:

- DSP-based plug-in motion controllers for PCI or PXI
- Custom motion controllers using the Schneider SoftMotion Module

3.2.1. Custom Motion Controllers

While plug-in motion controllers with DSPs are suitable for many applications, you need customized motion controllers for high-precision motion control with servo update rates as fast as 200 kHz. Some of the applications that require such a high level of precision and flexibility include wafer processing machines in the semiconductor industry or the inline vehicle sequencing (ILVS) reconfigurable-at-run-time assembly line for the automotive industry.

Schn

machine builders who want high-precision customized motion control with the flexibility of an FPGA. In addition to high-precision applications, machine builders and OEMs can use the Schneider SoftMotion Module to implement multiaxis coordinated motion control using the custom graphical development environment on a variety of platforms. For applications that require rugged and compact systems.

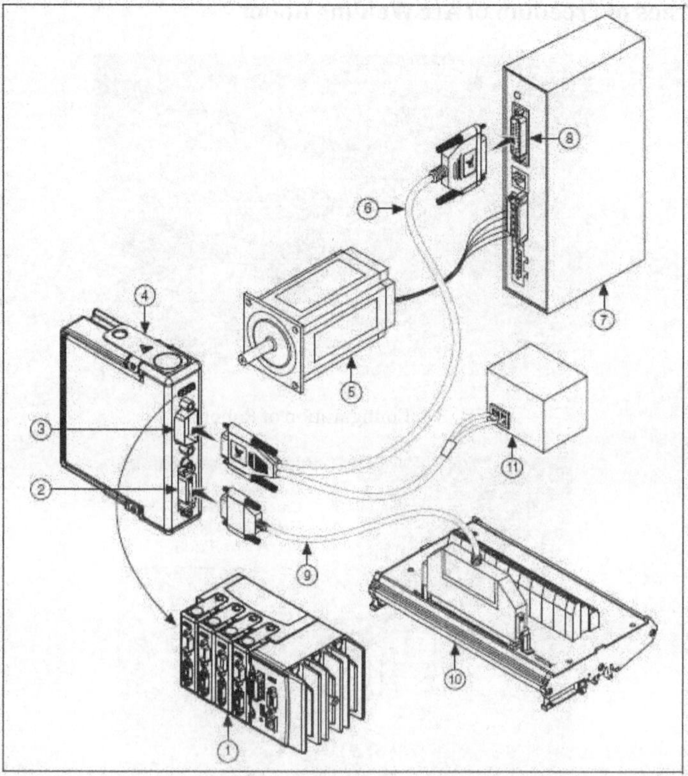

Fig. 3.3: Interfacing of Motion Controller

3.3. Designed Specification of Robot

Number of axis	-	Four
Load carrying capacity	-	40 kg
Maximum feed	-	5mm/sec
Maximum rapid	-	9000mm/min
Reach and Stroke	-	1940mm and 120mm
Tool orientation	-	perpendicular
Repeatability	-	±0.2mm
Precision and Accuracy	-	±0.3mm
Operating environment	-	Hazardous & HAZ

3.4. Degrees of Freedom of Arc Welding Robot

Fig. 3.4: Configuration of Robot

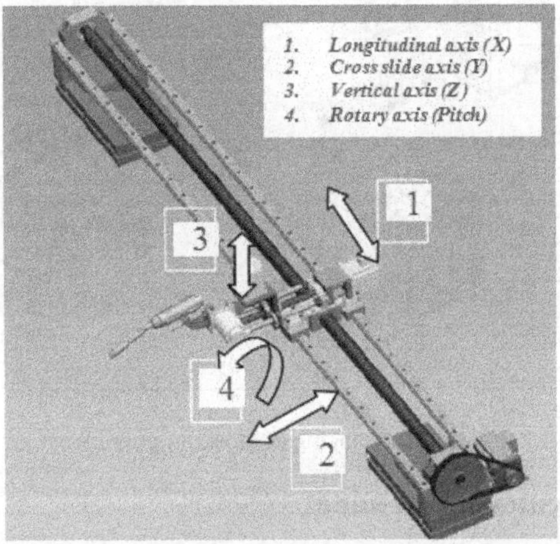

Fig. 3.5: Axis of Robot

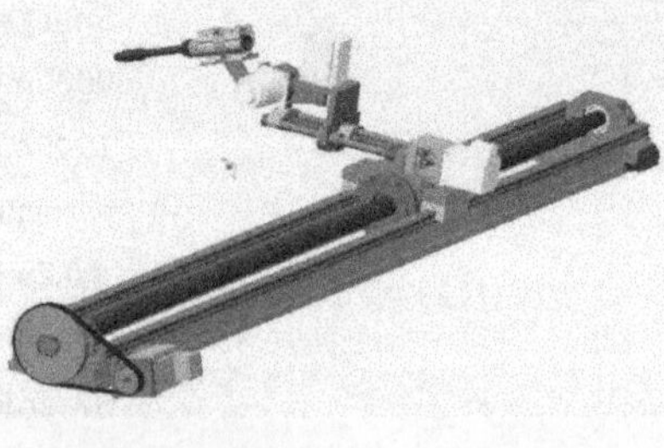

Fig. 3.6: Modeling View of Robot

CHAPTER 4

PARTS LIST

PARTS LIST				
MACHINE : 4-AXIS WELDING ROBOT				
Part No.	ITEM DESCRIPTION	MAKE	MODEL / DRG. NO.	QTY
01	END PLATE - I		0033-10-01	1 No
02	END PLATE - II		0033-10-02	1 No
03	SLIDE PLATE - I		0033-10-03	1 No
04	BALLSCREW - I	HIWIN	R32-10 T4 (Drg. No. 0033-10-04)	1 No
05	NUT HOUSING - I		0033-10-05	1 No
06	BALLSCREW PULLEY - I (Pitch - L (3/8"), No. of Teeth - 40)		0033-10-06	1 No
07	BEARING HOUSING - I		0033-10-07	1 No
08	BEARING HOUSING - II		0033-10-08	1 No
09	BEARING CAP - I		0033-10-09	1 No
10	BEARING CAP - II		0033-10-10	1 No
11	SERVO MOTOR PULLEY - I (Pitch - L (3/8"), No. of Teeth - 20)		0033-10-11	1 No
12	SERVO MOTOR MOUNTING PLATE - I		0033-10-12	1 No
13	SUPPORT RAIL BOTTOM PLATE		0033-10-13	6 Nos
14	LM SHAFT - Ø20 x 2000 L		0033-10-14	2 Nos
15	SHAFT SUPPORT RAIL		0033-10-15	2 Nos
16	LM BUSH BEARING WITH HOUSING	SAMICK	SBR20UU	4 Nos
17	AL. PROFILE BOTTOM BEAM	PBH	MPS 4080 H ; 2000 L	2 Nos
18	AL. PROFILE CROSS BEAM	PBH	MPS 4080 H ; 160 L	4 Nos
19	ANGLE BRACKET	PBH	DC/BK 8075	4 Nos
20	SPACER - I		0033-10-16	2 Nos
21	SPACER - II		0033-10-17	1 No
22	CLAMPING FLANGE - I		0033-10-18	1 No
23	CLAMPING FLANGE - II		0033-10-19	1 No
24	CHECK NUT		M25 x 1.5	2 Nos
25	T-NUT		0033-10-20	42 Nos
26	ANGULAR CONTACT BALL BEARING	FAG	7205	2 Nos
27	DEEP GROOVE BALL BEARING	FAG	6205	2 Nos
28	LOCKING ELEMENT	MAV	3003 - 20 x 25	2 Nos
29	LOCKING ELEMENT	MAV	3003 - 19 x 24	2 Nos
30	TIMING BELT	OPTIBELT	270L, 3/4" Width	1 No
31	SOC HEAD CAP SCREW		M8 x 40	8 Nos
32	SOC HEAD CAP SCREW		M8 x 25	6 Nos
33	SOC HEAD CAP SCREW		M8 x 25	4 Nos
34	SOC HEAD CAP SCREW		M6 x 25	16 Nos
35	SOC HEAD CAP SCREW		M6 x 25	12 Nos
36	SOC HEAD CAP SCREW		M6 x 16	52 Nos
37	SOC HEAD CAP SCREW		M5 x 12	42 Nos
38	EXTERNAL CIRCLIP		A25	1 No

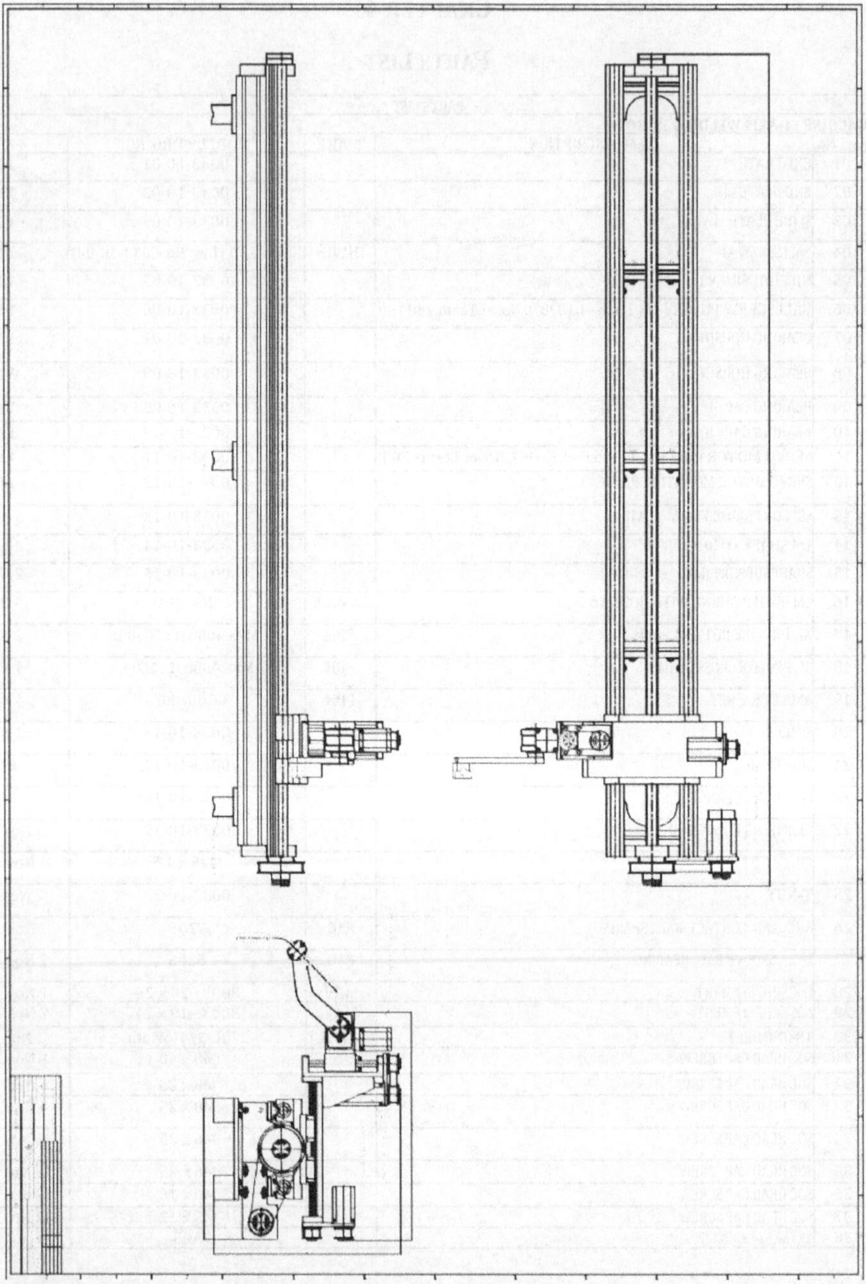

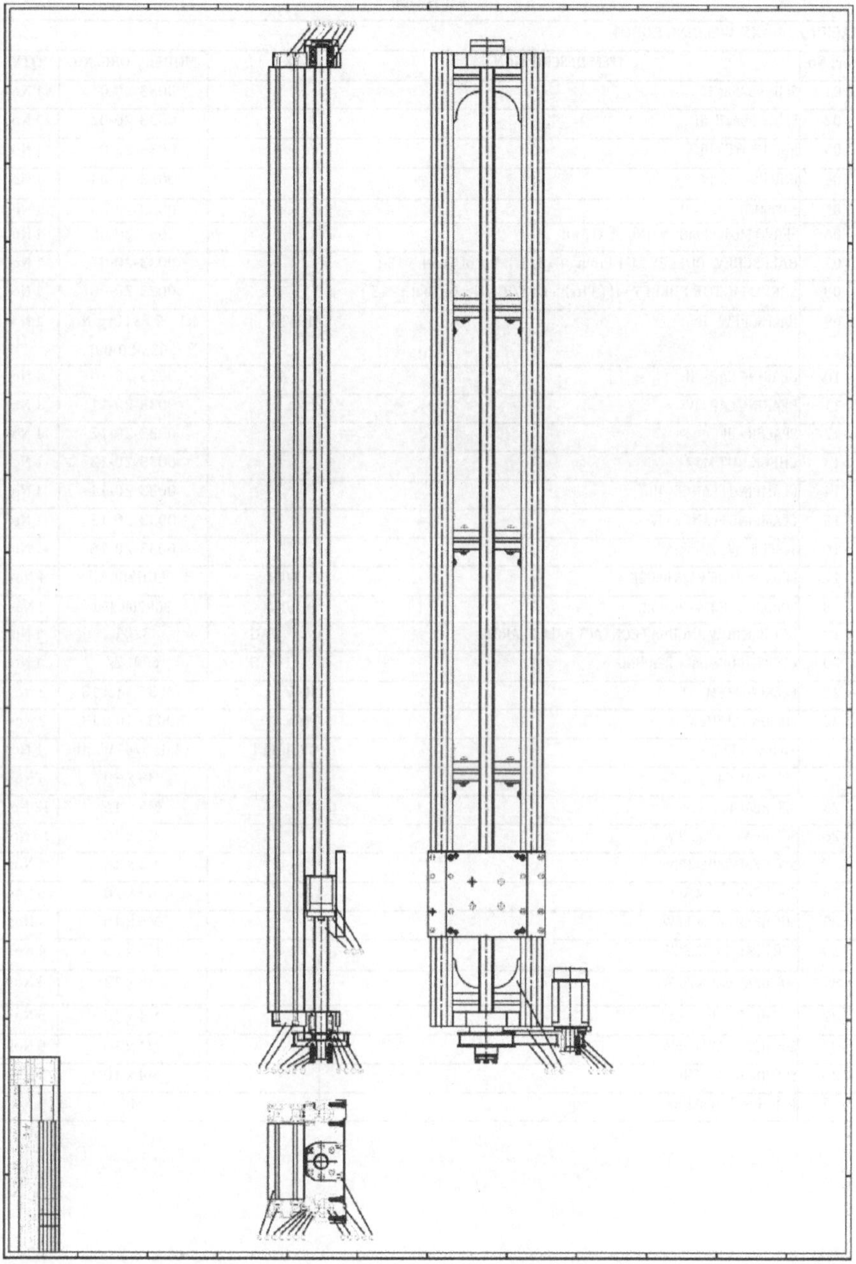

PARTS LIST				
MACHINE : 4-AXIS WELDING ROBOT				
Part No.	**ITEM DESCRIPTION**	**MAKE**	**MODEL / DRG. NO.**	**QTY**
01	SLIDE BASE - I		0033-20-01	1 No
02	SLIDE PLATE - II		0033-20-02	1 No
03	END PLATE - III		0033-20-03	1 No
04	END PLATE - IV		0033-20-04	1 No
05	NUT HOUSING - II		0033-20-05	1 No
06	SERVO MOTOR MOUNTING PLATE - II		0033-20-06	1 No
07	BALLSCREW PULLEY - II (Pitch - L (3/8"), No. of Teeth - 15)		0033-20-07	1 No
08	SERVO MOTOR PULLEY - II (Pitch - L (3/8"), No. of Teeth - 15)		0033-20-08	1 No
09	BALLSCREW - II	HIWIN	R16-5 T3 (Drg. No. 0033-20-09)	1 No
10	BEARING CAP - III		0033-20-10	1 No
11	BEARING CAP - IV		0033-20-11	1 No
12	SPACER - III		0033-20-12	1 No
13	CHECK NUT M12		0033-20-13	1 No
14	CLAMPING FLANGE - III		0033-20-14	1 No
15	CLAMPING FLANGE - IV		0033-20-15	1 No
16	SPACER - IV		0033-20-16	1 No
17	LINEAR GUIDEWAYS BLOCK	HIWIN	EGH20CA	4 Nos
18	LINEAR GUIDEWAYS RAIL	HIWIN	EGR20R 350	2 Nos
19	DOUBLE ROW ANGULAR CONTACT BALL BEARING	FAG	3201	1 No
20	DEEP GROOVE BALL BEARING	FAG	6001 ZZ	1 No
21	LOCKING ELEMENT	MAV	3003 - 14 x 18	2 Nos
22	LOCKING ELEMENT	MAV	3003 - 10 x 13	2 Nos
23	TIMING BELT	OPTIBELT	124L, 3/4" Width	1 No
24	SOC HEAD CAP SCREW		M8 x 30	4 Nos
25	SOC HEAD CAP SCREW		M5 x 16	12 Nos
26	SOC HEAD CAP SCREW		M5 x 16	16 Nos
27	SOC HEAD CAP SCREW		M5 x 16	4 Nos
28	SOC HEAD CAP SCREW		M5 x 20	6 Nos
29	SOC HEAD CAP SCREW		M5 x 16	4 Nos
30	SOC HEAD CAP SCREW		M6 x 25	4 Nos
31	SOC HEAD CAP SCREW		M5 x 12	4 Nos
32	SOC HEAD CAP SCREW		M4 x 16	6 Nos
33	SOC HEAD CAP SCREW		M4 x 16	6 Nos
34	SOC HEAD CSK SCREW		M4 x 10	3 Nos
35	MACHINED WASHER		M6	4 Nos

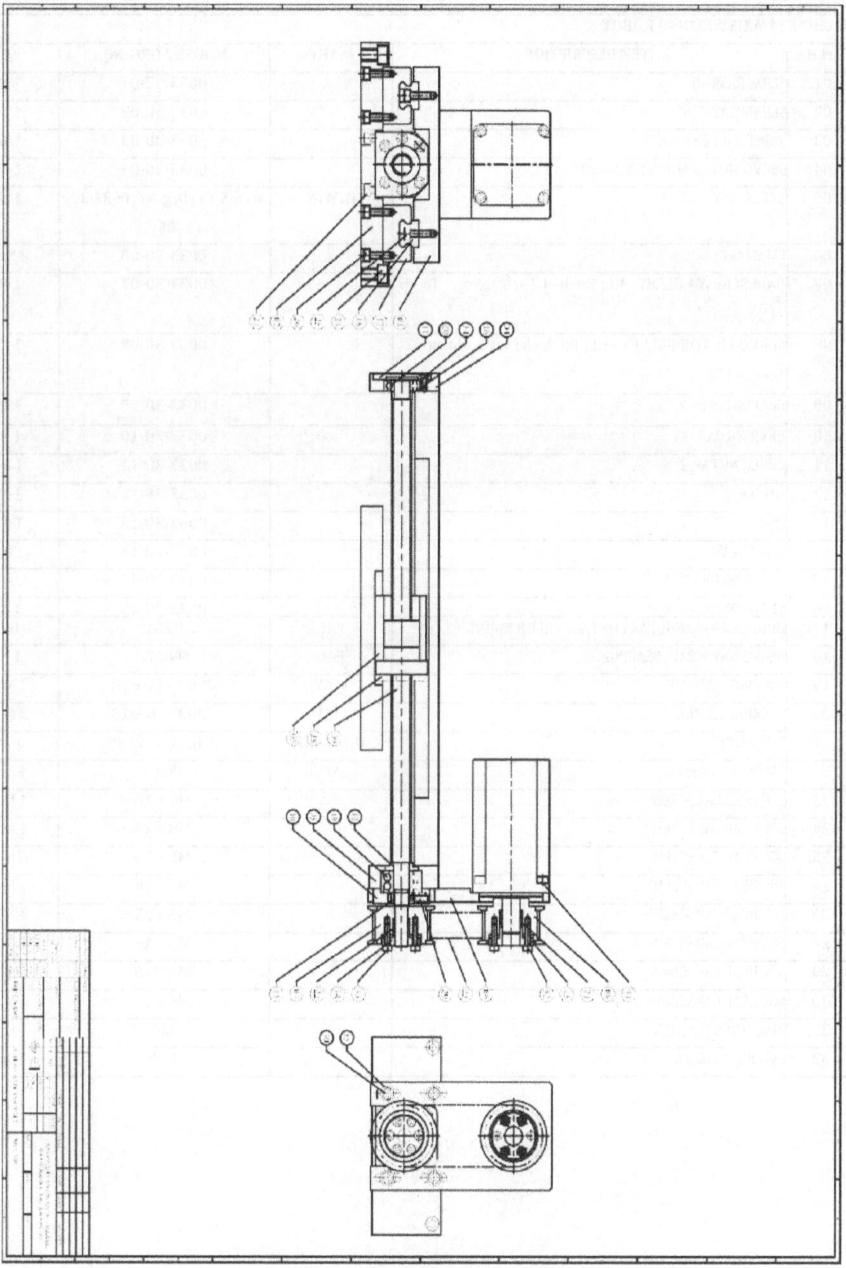

PARTS LIST				
MACHINE : 4-AXIS WELDING ROBOT				
Part No.	**ITEM DESCRIPTION**	**MAKE**	**MODEL / DRG. NO.**	**QTY**
01	SLIDE BASE - II		0033-30-01	1 No
02	SLIDE PLATE - III		0033-30-02	1 No
03	RIGHT ANGLE BRACKET		0033-30-03	1 No
04	SERVO MOTOR MOUNTING PLATE - III		0033-30-04	1 No
05	BALLSCREW - III	HIWIN	R16-5 T3 (Drg. No. 0033-30-05)	1 No
06	LM SHAFT		0033-30-06	2 Nos
07	BALLSCREW PULLEY - III (Pitch - L (3/8"), No. of Teeth - 15)		0033-30-07	1 No
08	SERVO MOTOR PULLEY - III (Pitch - L (3/8"), No. of Teeth - 15)		0033-30-08	1 No
09	BEARING CAP - V		0033-30-09	1 No
10	BEARING CAP - VI		0033-30-10	1 No
11	CHECK NUT M12		0033-30-11	1 No
12	SPACER - V		0033-30-12	1 No
13	SPACER - VI		0033-30-13	1 No
14	SPACER - VII		0033-30-14	1 No
15	CLAMPING FLANGE - V		0033-30-15	1 No
16	CLAMPING FLANGE - VI		0033-30-16	1 No
17	DOUBLE ROW ANGULAR CONTACT BALL BEARING	FAG	3201	1 No
18	DEEP GROOVE BALL BEARING	FAG	6001 ZZ	1 No
19	LOCKING ELEMENT	MAV	3003 - 14 x 18	2 Nos
20	LOCKING ELEMENT	MAV	3003 - 10 x 13	2 Nos
21	TIMING BELT	OPTIBELT	124L, 3/4" Width	1 No
22	LM BUSH BEARING	SAMICK	LME16 UU	4 Nos
23	SOC HEAD CAP SCREW		M8 x 16	4 Nos
24	SOC HEAD CAP SCREW		M8 x 25	4 Nos
25	SOC HEAD CAP SCREW		M5 x 20	6 Nos
26	SOC HEAD CAP SCREW		M5 x 16	4 Nos
27	SOC HEAD CAP SCREW		M6 x 25	4 Nos
28	SOC HEAD CAP SCREW		M4 x 16	6 Nos
29	SOC HEAD CAP SCREW		M4 x 16	6 Nos
30	SOC HEAD CAP SCREW		M5 x 12	4 Nos
31	MACHINED WASHER		M6	4 Nos
32	INTERNAL CIRCLIP		B26	4 Nos

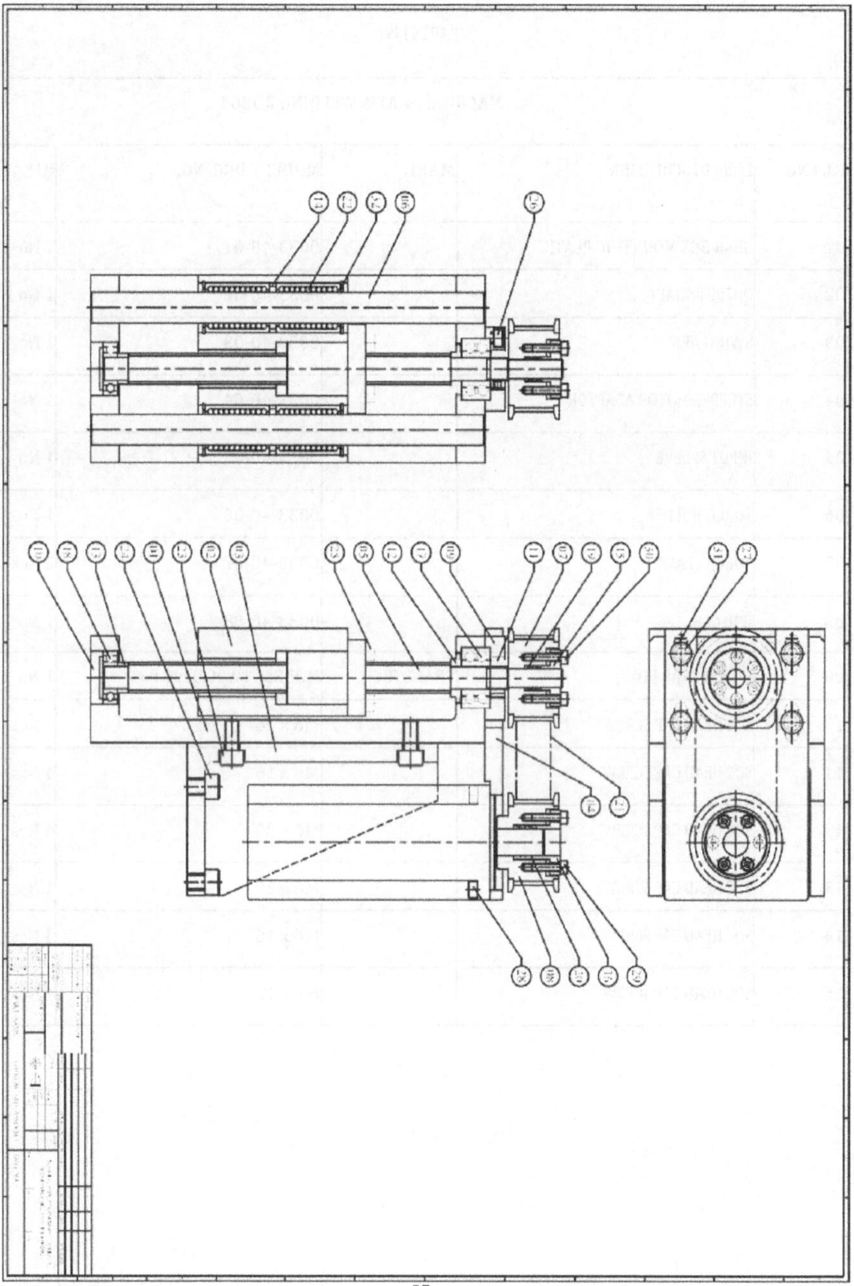

PARTS LIST				
MACHINE : 4-AXIS WELDING ROBOT				
Part No.	**ITEM DESCRIPTION**	**MAKE**	**MODEL / DRG. NO.**	**QTY**
01	GEAR BOX MOUNTING PLATE		0033-40-01	1 No
02	OUTPUT SHAFT		0033-40-02	1 No
03	SWING ARM		0033-40-03	1 No
04	STEPPER MOTOR ADAPTOR		0033-40-04	1 No
05	INPUT SLEEVE		0033-40-05	1 No
06	TORCH HOLDER		0033-40-06	1 No
07	TORCH CLAMP		0033-40-07	2 Nos
08	END CAP		0033-40-08	1 No
09	WORM GEAR BOX	TRANSTECH	MODEL - TW30 U100 F63	1 No
10	SOC HEAD CAP SCREW		M6 x 16	4 Nos
11	SOC HEAD CAP SCREW		M6 x 16	4 Nos
12	SOC HEAD CAP SCREW		M6 x 20	4 Nos
13	SOC HEAD CAP SCREW		M6 x 20	1 No
14	SOC HEAD CAP SCREW		M6 x 16	4 Nos
15	SOC HEAD CAP SCREW		M4 x 12	4 Nos

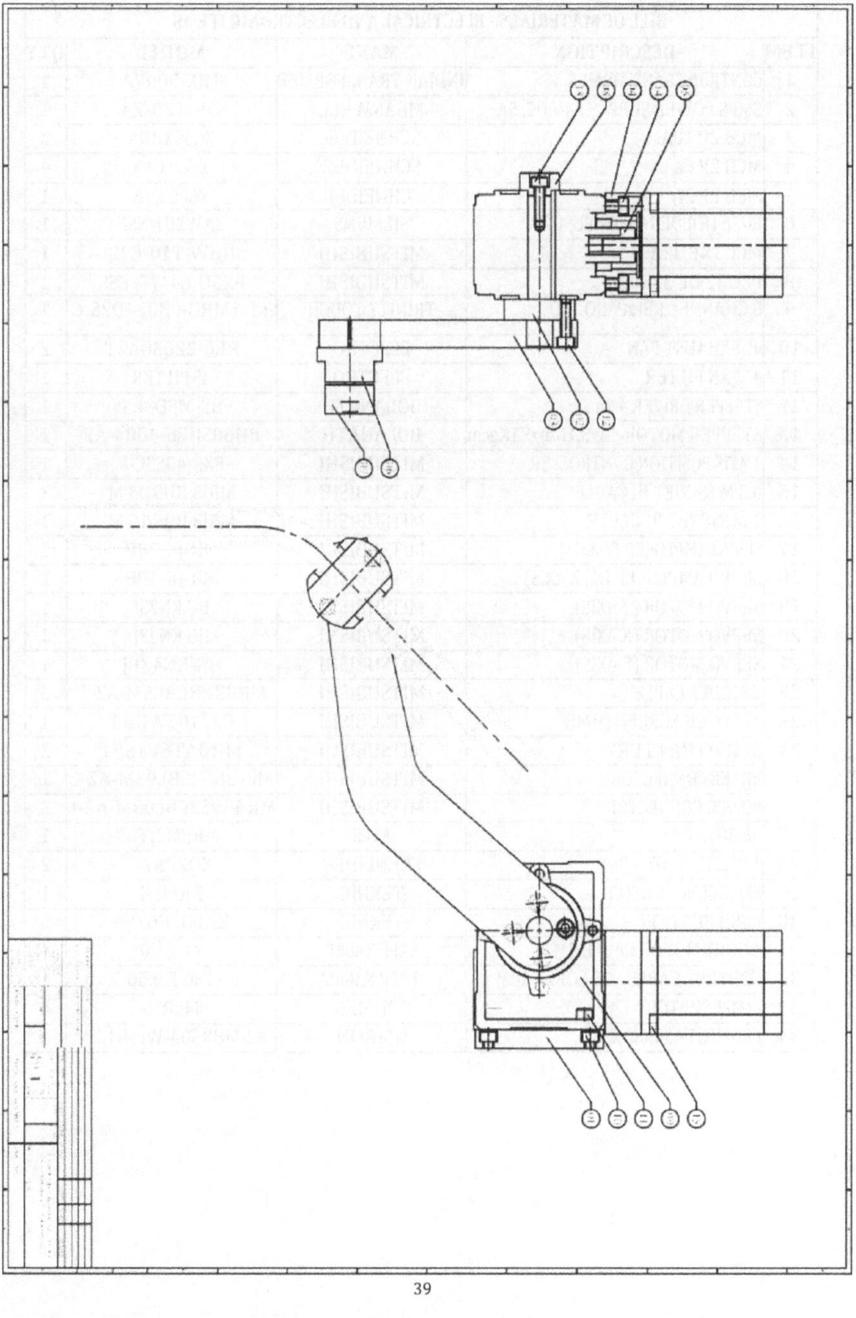

BILL OF MATERIALS - ELECTRICAL AND ELECTRONIC ITEMS				
ITEM	DESCRIPTION	MAKE	MODEL	QTY
1	CONTROL TRANSFORMER	INDIAN TRANSFORMER	TRX-500VA	1
2	SMPS POWER SUPPLY 24V DC, 5A	MEANWELL	DR-120-24	1
3	MCB 2P 10A	SCHNEIDER	XC60 C10A	3
4	MCB 2P 6A	SCHNEIDER	XC60 C6A	4
5	MCB 2P 2A	SCHNEIDER	XC60 C2A	1
6	FUSE HOLDER WITH FUSE 4A	SIEMENS	3NW20 NNSF	1
7	MCCB 4P 32A	MITSUBISHI	BHW-T10-C32	1
8	PLC 32 DI, 32DO	MITSUBISHI	FX5U-64MT/ESS	1
9	8 CHANNEL RELAY BOARD	TRINITY TOUCH	TT-1MRB-08024D2S-C	1
10	4" EXHAUST FAN	ELECTRO	REC-22038A2T	2
11	4" FAN FILTER	ELECTRO	4" FILTER	1
12	STEPPER DRIVER 4.5A	BOLANATH	BS-MSD-4.5	1
13	STEPPER MOTOR 4A, 3.1Nm/31Kgcm	BOLANATH	BH60SH86-4004-AF	1
14	4 AXIS POSITION CONTROLLER	MITSUBISHI	FX5-40SSC-S	1
15	0.3 M SSCNET III CABLE	MITSUBISHI	MRJ3 BUS 03 M	3
16	1 M SSCNET III CABLE	MITSUBISHI	MRJ3 BUS 10 M	1
17	SERVO AMPLIFIER (Y AXIS)	MITSUBISHI	MR-JE-70B	1
18	SERVO AMPLIFIER (X , Z AXIS)	MITSUBISHI	MR-JE-40B	2
19	SERVO MOTOR (Y AXIS)	MITSUBISHI	HG KN73J	1
20	SERVO MOTOR (X AXIS)	MITSUBISHI	HG KN43J	1
21	SERVO MOTOR (Z AXIS)	MITSUBISHI	HG-KN43BJ	1
22	ENCODER CABLE	MITSUBISHI	MR-J3ENCBL5M-A2-L	3
23	7" TOUCH SCREEN (HMI)	MITSUBISHI	GS2107WBDA	1
24	SERVO BATTERY	MITSUBISHI	MRBAT6V1SET	2
25	BRAKE CONNECTOR	MITSUBISHI	MR-BKS2CBL03M-A2-L	1
26	POWER CONNECTOR	MITSUBISHI	MR-PWS2CBL03M-A2-L	3
27	REMOTE BOX	AWS	REM2JY	1
28	4 POSITION JOY STICK	SCHNEIDER	XD2PA24	2
29	EMERGENCY SWITCH	TEKNIC	2AML4	1
30	PUSH BUTTION	TEKNIC	2AF8/F6/F4	5
31	4 CORE MOTOR CABLE 8M	LAPP KABEL	4G X1.0	4
32	CONTROL CABLE - 18 CORE - 10M	LAPP KABEL	18G X 0.50	1
33	LIMIT SWITCH	BOHMEN	MLRLS	4
34	PROXIMITY SENSOR	OMRON	E2-M8KS04-WP-B1 2M	2

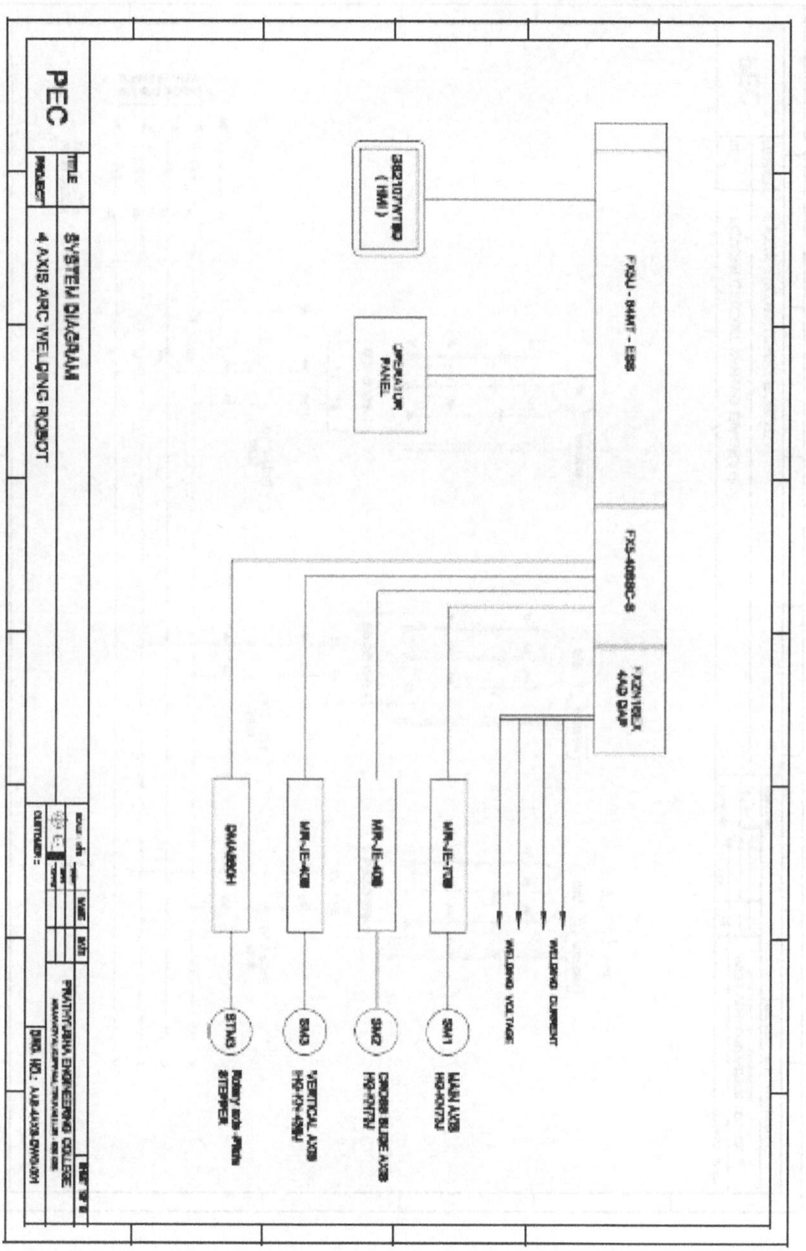
PEC
TITLE
SYSTEM DIAGRAM
PROJECT
4 AXIS ARC WELDING ROBOT
OPERATOR PANEL
WELDING CURRENT
WELDING VOLTAGE
SM1
SM2
SM3
STM3
MAIN AXIS
CROSS SLIDE AXIS
VERTICAL AXIS
STEPPER
CUSTOMER :

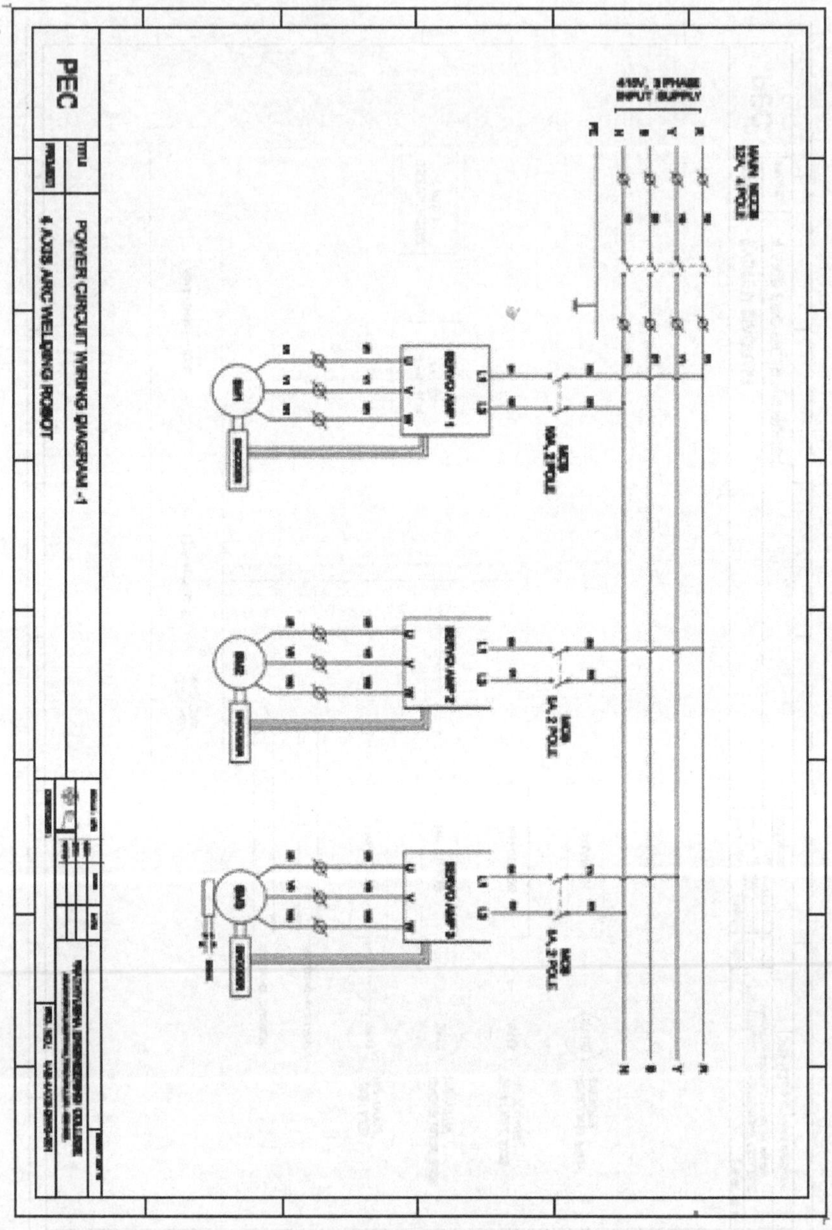
PEC
POWER CIRCUIT WIRING DIAGRAM -1
4 AXIS ARC WELDING ROBOT
415V, 3 PHASE INPUT SUPPLY
SERVO AMP 1
SERVO AMP 2
SERVO AMP 3
ENCODER

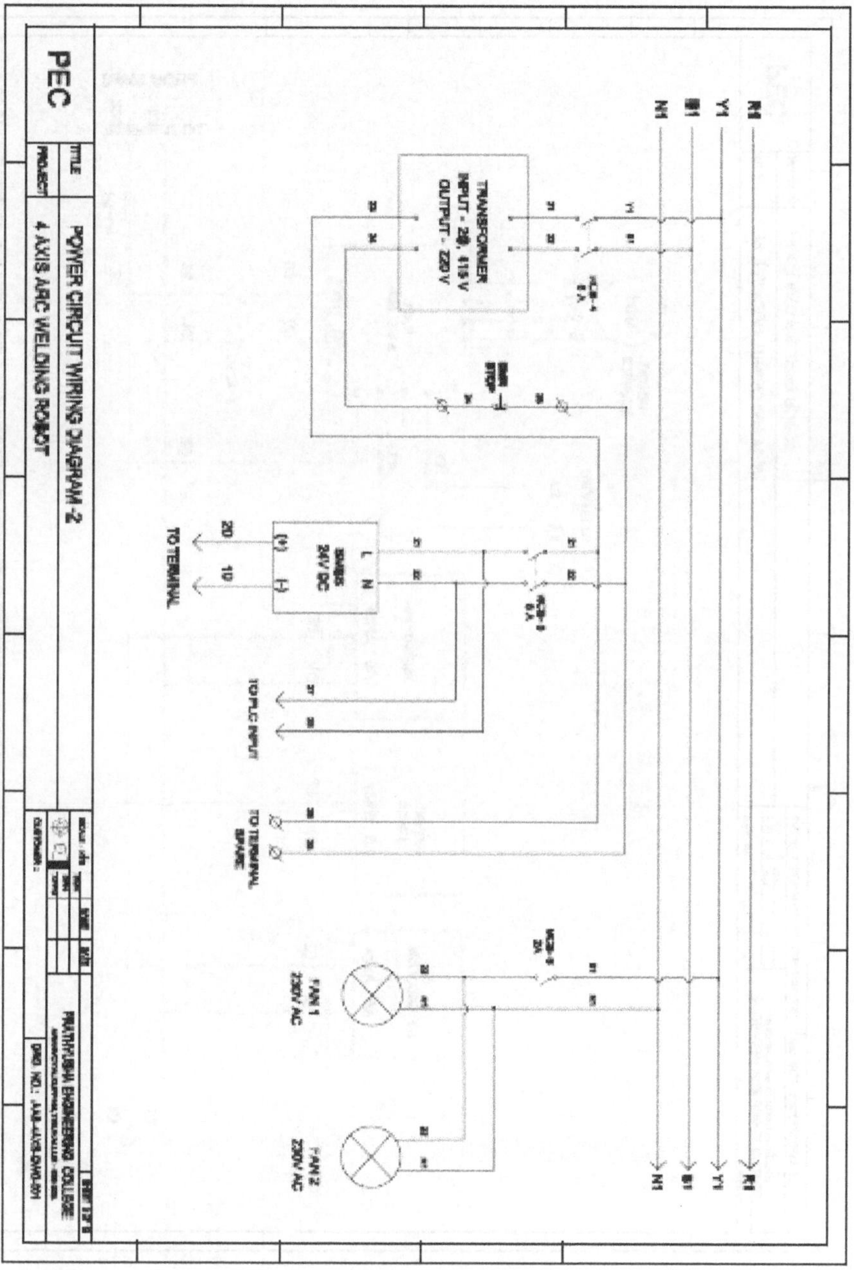

PEC
TITLE
POWER CIRCUIT WIRING DIAGRAM -2
PROJECT
4 AXIS ARC WELDING ROBOT
R1
Y1
B1
N1
TRANSFORMER
INPUT - 2Φ, 415 V
OUTPUT - 220V
SMPS
24V DC
L
N
(+)
(-)
20
10
TO TERMINAL
TO PLC INPUT
TO TERMINAL
SPARE
FAN 1
230V AC
FAN 2
230V AC
PRATHYUSHA ENGINEERING COLLEGE

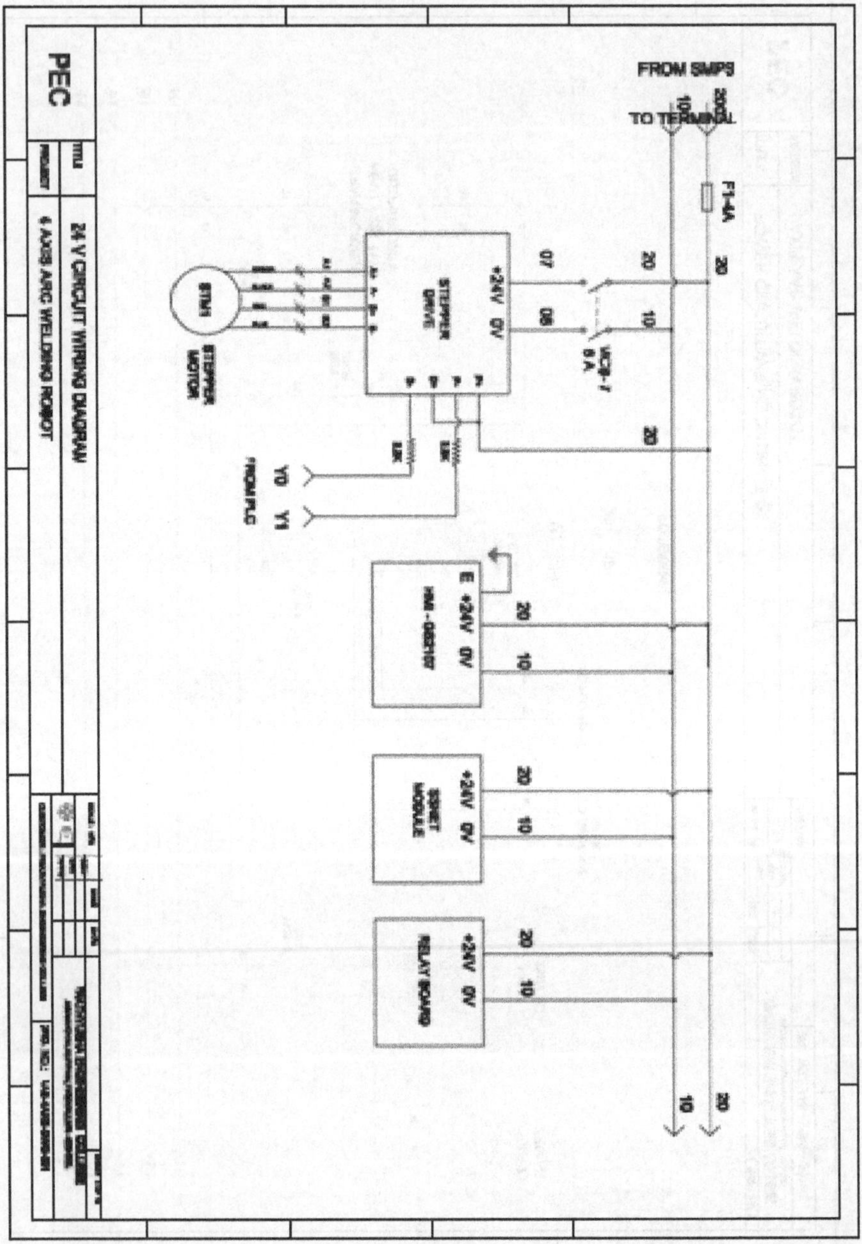

FROM SMPS
TO TERMINAL
F1-4A
STEPPER DRIVE
+24V
0V
STEPPER MOTOR
FROM PLC
Y0
Y1
RELAY BOARD
PEC
24 V CIRCUIT WIRING DIAGRAM

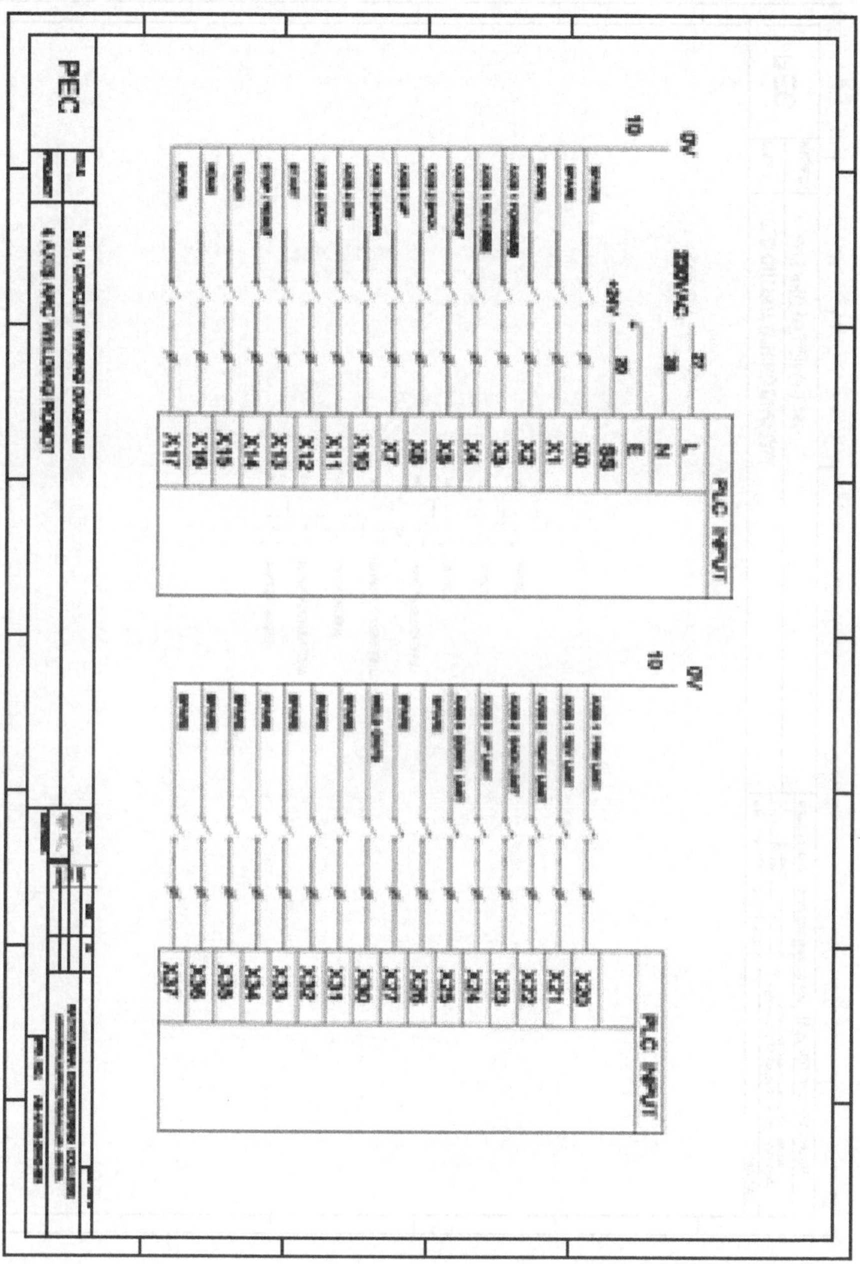
PEC
24 V CIRCUIT WIRING DIAGRAM
4 AXIS ARC WELDING ROBOT
PLC INPUT
0V
10
230VAC
+24V
L
N
E
X0
X1
X2
X3
X4
X5
X6
X7
X10
X11
X12
X13
X14
X15
X16
X17
PLC INPUT
0V
10
X20
X21
X22
X23
X24
X25
X26
X27
X30
X31
X32
X33
X34
X35
X36
X37

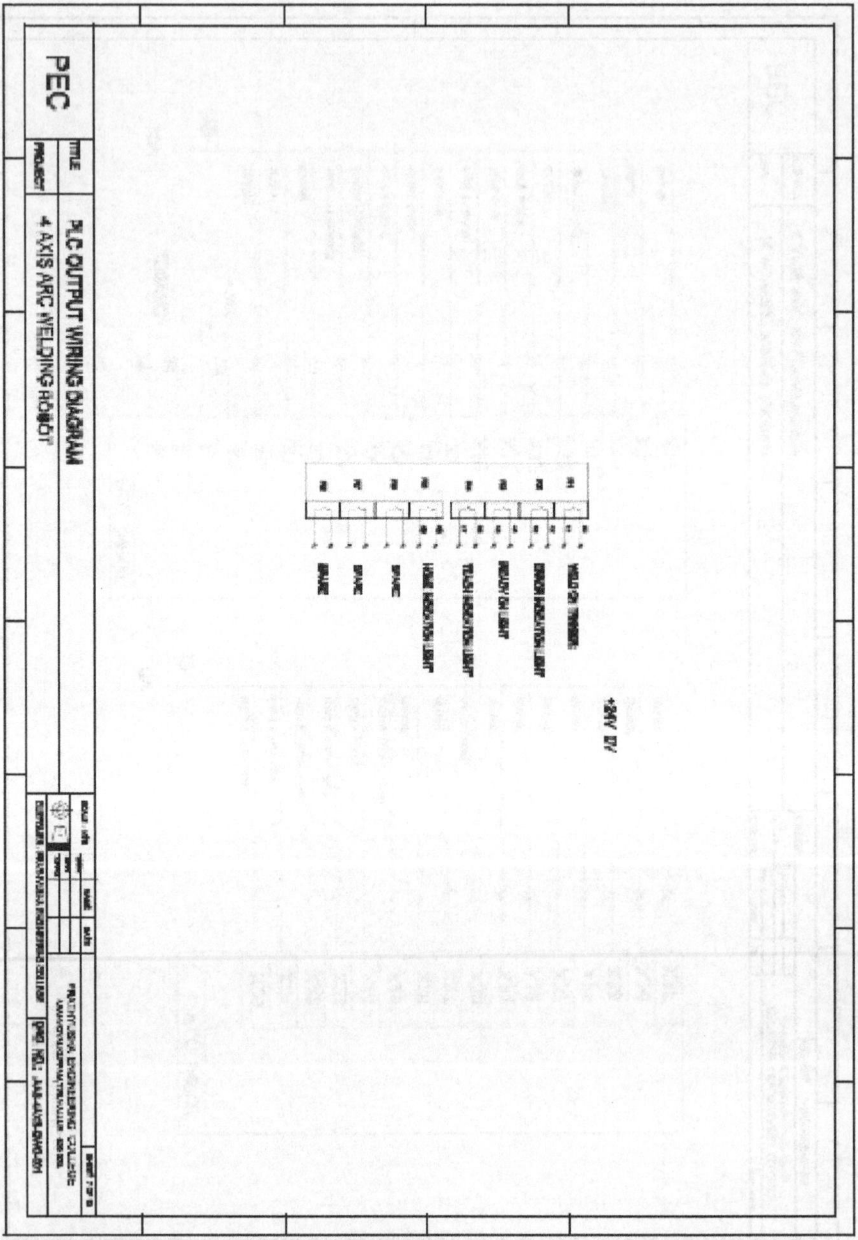

+24V 0V
WELD ON TRIGGER
ERROR INDICATION LIGHT
READY ON LIGHT
TEACH INDICATION LIGHT
HOME INDICATION LIGHT
SPARE
SPARE
SPARE
PEC
TITLE
PLC OUTPUT WIRING DIAGRAM
PROJECT
4 AXIS ARC WELDING ROBOT
PRATHYUSHA ENGINEERING COLLEGE

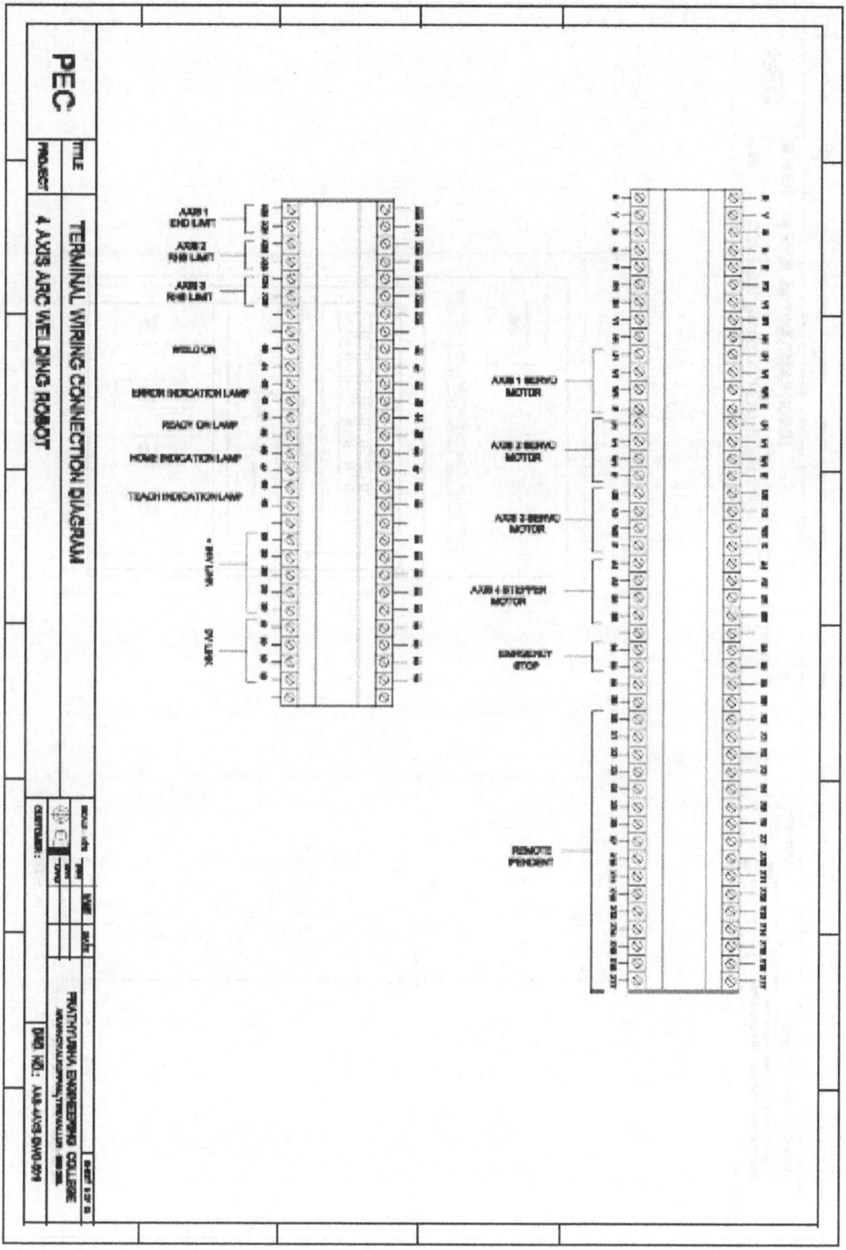
AXIS 1 SERVO MOTOR
AXIS 2 SERVO MOTOR
AXIS 3 SERVO MOTOR
AXIS 4 STEPPER MOTOR
EMERGENCY STOP
REMOTE PENDENT
AXIS 1 END LIMIT
AXIS 2 RHS LIMIT
AXIS 3 RHS LIMIT
WELD ON
ERROR INDICATION LAMP
READY ON LAMP
HOME INDICATION LAMP
TEACH INDICATION LAMP
0V LINK
PEC
TITLE
TERMINAL WIRING CONNECTION DIAGRAM
PROJECT
4 AXIS ARC WELDING ROBOT
CUSTOMER :
PRATHYUSHA ENGINEERING COLLEGE

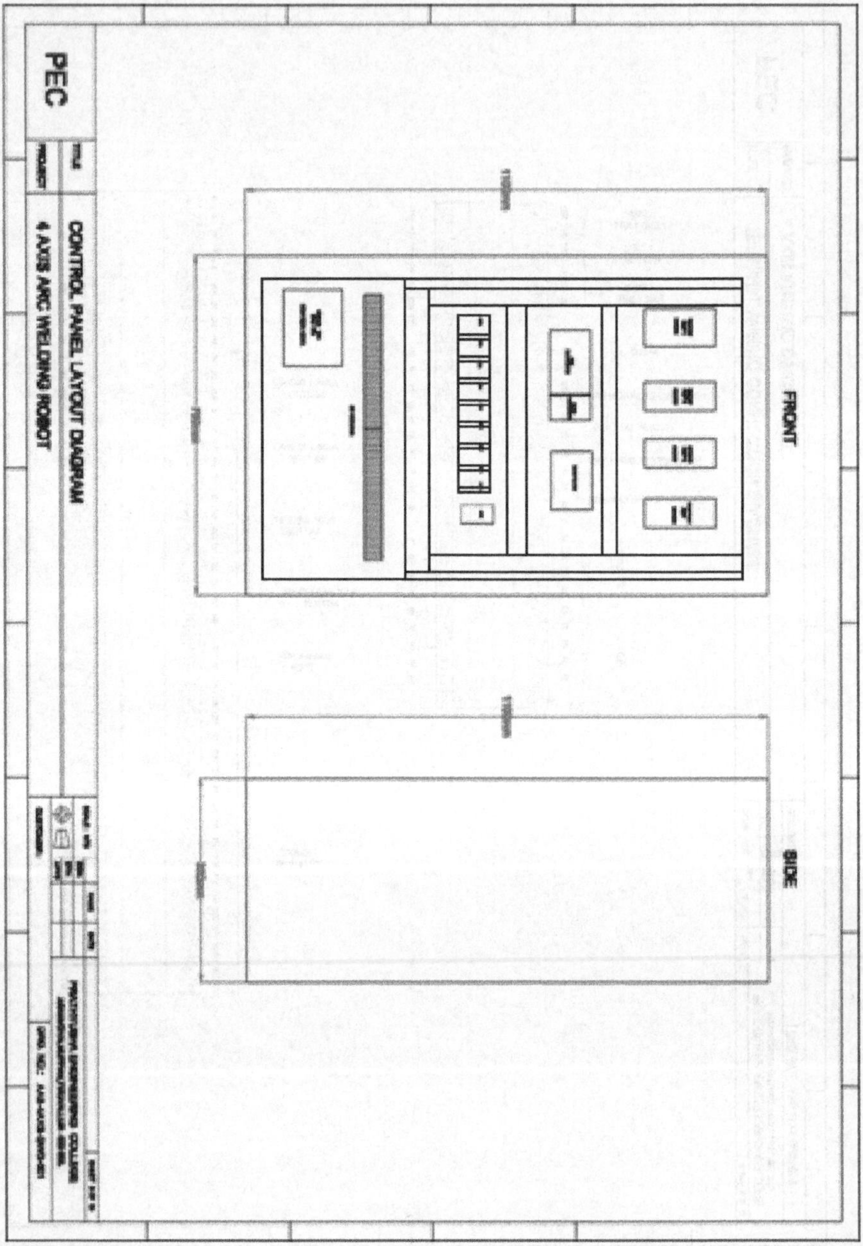
PEC
TITLE
CONTROL PANEL LAYOUT DIAGRAM
PROJECT
4 AXIS ARC WELDING ROBOT
FRONT
SIDE

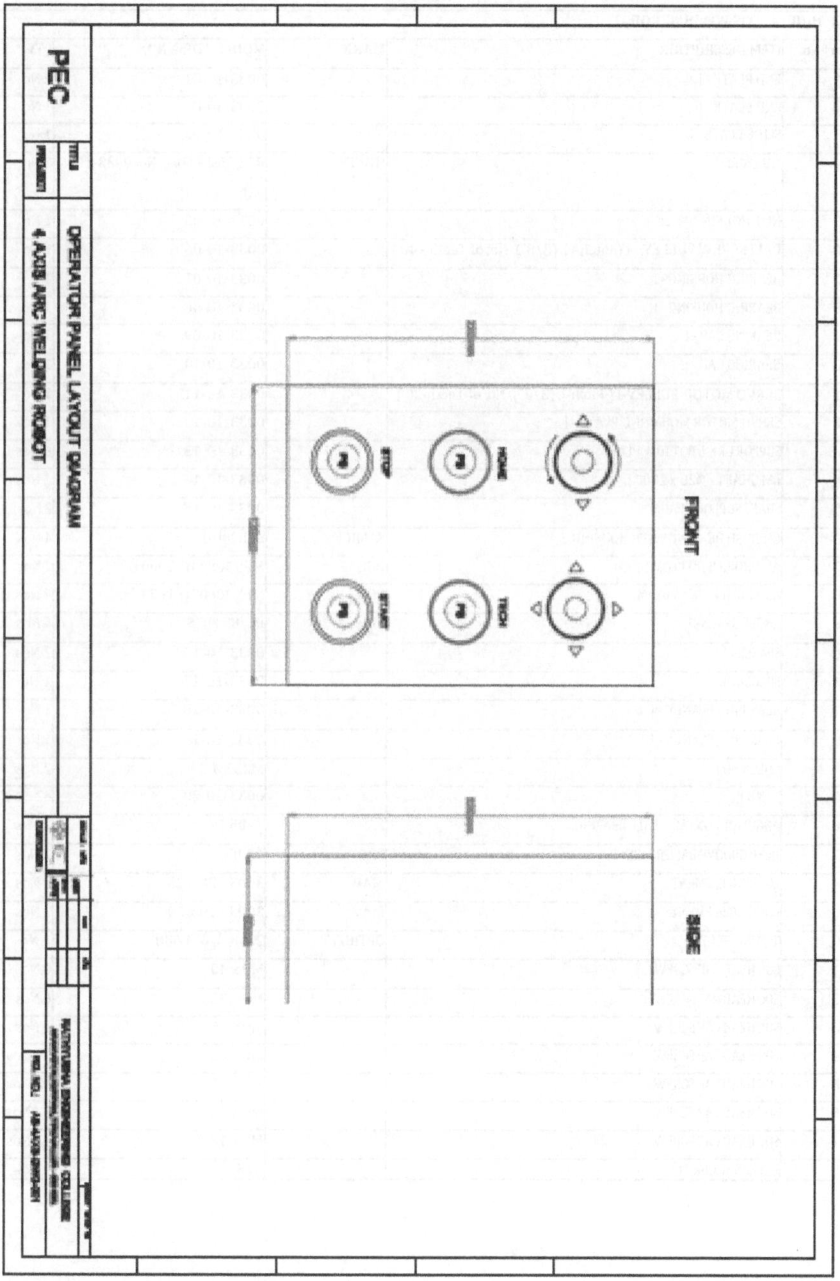
PEC
TITLE
OPERATOR PANEL LAYOUT DIAGRAM
PROJECT
4 AXIS ARC WELDING ROBOT
FRONT
HOME
STOP
TECH
START
SIDE

PARTS LIST				
MACHINE : 4-AXIS WELDING ROBOT				
Part No.	**ITEM DESCRIPTION**	**MAKE**	**MODEL / DRG. NO.**	**QTY**
01	END PLATE - I		0033-10-01	1 No
02	END PLATE - II		0033-10-02	1 No
03	SLIDE PLATE - I		0033-10-03	1 No
04	BALLSCREW - I	HIWIN	R32-10 T4 (Drg. No. 0033-10-04)	1 No
05	NUT HOUSING - I		0033-10-05	1 No
06	BALLSCREW PULLEY - I (Pitch - L (3/8"), No. of Teeth - 40)		0033-10-06	1 No
07	BEARING HOUSING - I		0033-10-07	1 No
08	BEARING HOUSING - II		0033-10-08	1 No
09	BEARING CAP - I		0033-10-09	1 No
10	BEARING CAP - II		0033-10-10	1 No
11	SERVO MOTOR PULLEY-I (Pitch-L (3/8"), No. of Teeth-20)		0033-10-11	1 No
12	SERVO MOTOR MOUNTING PLATE - I		0033-10-12	1 No
13	SUPPORT RAIL BOTTOM PLATE		0033-10-13	6 Nos
14	LM SHAFT - Ø20 x 2000 L		0033-10-14	2 Nos
15	SHAFT SUPPORT RAIL		0033-10-15	2 Nos
16	LM BUSH BEARING WITH HOUSING	SAMICK	SBR20UU	4 Nos
17	AL. PROFILE BOTTOM BEAM	PBH	MPS 4080 H ; 2000 L	2 Nos
18	AL. PROFILE CROSS BEAM	PBH	MPS 4080 H ; 160 L	4 Nos
19	ANGLE BRACKET	PBH	DC/BK 8075	4 Nos
20	SPACER - I		0033-10-16	2 Nos
21	SPACER - II		0033-10-17	1 No
22	CLAMPING FLANGE - I		0033-10-18	1 No
23	CLAMPING FLANGE - II		0033-10-19	1 No
24	CHECK NUT		M25 x 1.5	2 Nos
25	T-NUT		0033-10-20	42 Nos
26	ANGULAR CONTACT BALL BEARING	FAG	7205	2 Nos
27	DEEP GROOVE BALL BEARING	FAG	6205	2 Nos
28	LOCKING ELEMENT	MAV	3003 - 20 x 25	2 Nos
29	LOCKING ELEMENT	MAV	3003 - 19 x 24	2 Nos
30	TIMING BELT	OPTIBELT	270L, 3/4" Width	1 No
31	SOC HEAD CAP SCREW		M8 x 40	8 Nos
32	SOC HEAD CAP SCREW		M8 x 25	6 Nos
33	SOC HEAD CAP SCREW		M8 x 25	4 Nos
34	SOC HEAD CAP SCREW		M6 x 25	16 Nos
35	SOC HEAD CAP SCREW		M6 x 25	12 Nos
36	SOC HEAD CAP SCREW		M6 x 16	52 Nos
37	SOC HEAD CAP SCREW		M5 x 12	42 Nos
38	EXTERNAL CIRCLIP		A25	1 No

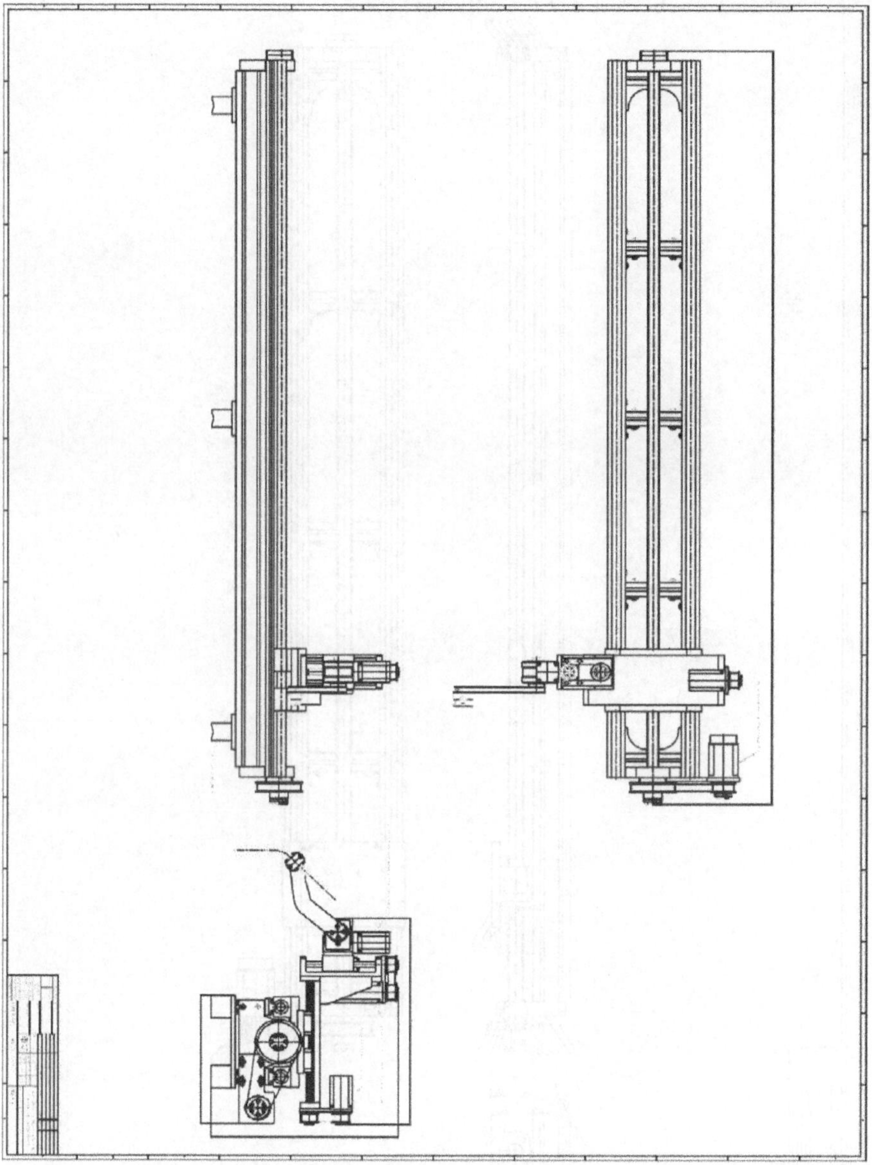

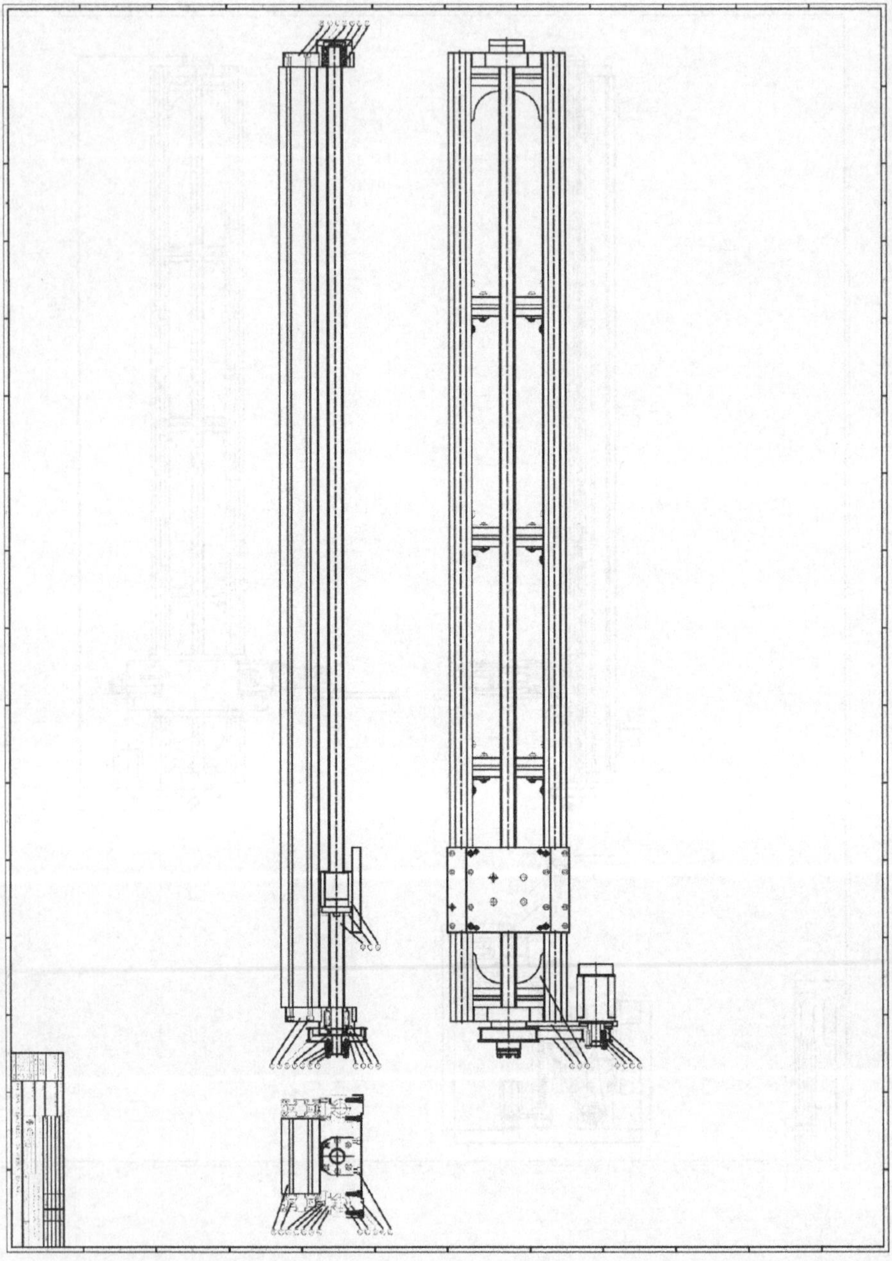

PARTS LIST				
MACHINE : 4-AXIS WELDING ROBOT				
Part No.	**ITEM DESCRIPTION**	**MAKE**	**MODEL / DRG. NO.**	**QTY**
01	SLIDE BASE - I		0033-20-01	1 No
02	SLIDE PLATE - II		0033-20-02	1 No
03	END PLATE - III		0033-20-03	1 No
04	END PLATE - IV		0033-20-04	1 No
05	NUT HOUSING - II		0033-20-05	1 No
06	SERVO MOTOR MOUNTING PLATE - II		0033-20-06	1 No
07	BALLSCREW PULLEY - II (Pitch - L (3/8"), No. of Teeth - 15)		0033-20-07	1 No
08	SERVO MOTOR PULLEY - II (Pitch - L (3/8"), No. of Teeth - 15)		0033-20-08	1 No
09	BALLSCREW - II	HIWIN	R16-5 T3 (Drg. No. 0033-20-09)	1 No
10	BEARING CAP - III		0033-20-10	1 No
11	BEARING CAP - IV		0033-20-11	1 No
12	SPACER - III		0033-20-12	1 No
13	CHECK NUT M12		0033-20-13	1 No
14	CLAMPING FLANGE - III		0033-20-14	1 No
15	CLAMPING FLANGE - IV		0033-20-15	1 No
16	SPACER - IV		0033-20-16	1 No
17	LINEAR GUIDEWAYS BLOCK	HIWIN	EGH20CA	4 Nos
18	LINEAR GUIDEWAYS RAIL	HIWIN	EGR20R 350	2 Nos
19	DOUBLE ROW ANGULAR CONTACT BALL BEARING	FAG	3201	1 No
20	DEEP GROOVE BALL BEARING	FAG	6001 ZZ	1 No
21	LOCKING ELEMENT	MAV	3003 - 14 x 18	2 Nos
22	LOCKING ELEMENT	MAV	3003 - 10 x 13	2 Nos
23	TIMING BELT	OPTIBELT	124L, 3/4" Width	1 No
24	SOC HEAD CAP SCREW		M8 x 30	4 Nos
25	SOC HEAD CAP SCREW		M5 x 16	12 Nos
26	SOC HEAD CAP SCREW		M5 x 16	16 Nos
27	SOC HEAD CAP SCREW		M5 x 16	4 Nos
28	SOC HEAD CAP SCREW		M5 x 20	6 Nos
29	SOC HEAD CAP SCREW		M5 x 16	4 Nos
30	SOC HEAD CAP SCREW		M6 x 25	4 Nos
31	SOC HEAD CAP SCREW		M5 x 12	4 Nos
32	SOC HEAD CAP SCREW		M4 x 16	6 Nos
33	SOC HEAD CAP SCREW		M4 x 16	6 Nos
34	SOC HEAD CSK SCREW		M4 x 10	3 Nos
35	MACHINED WASHER		M6	4 Nos

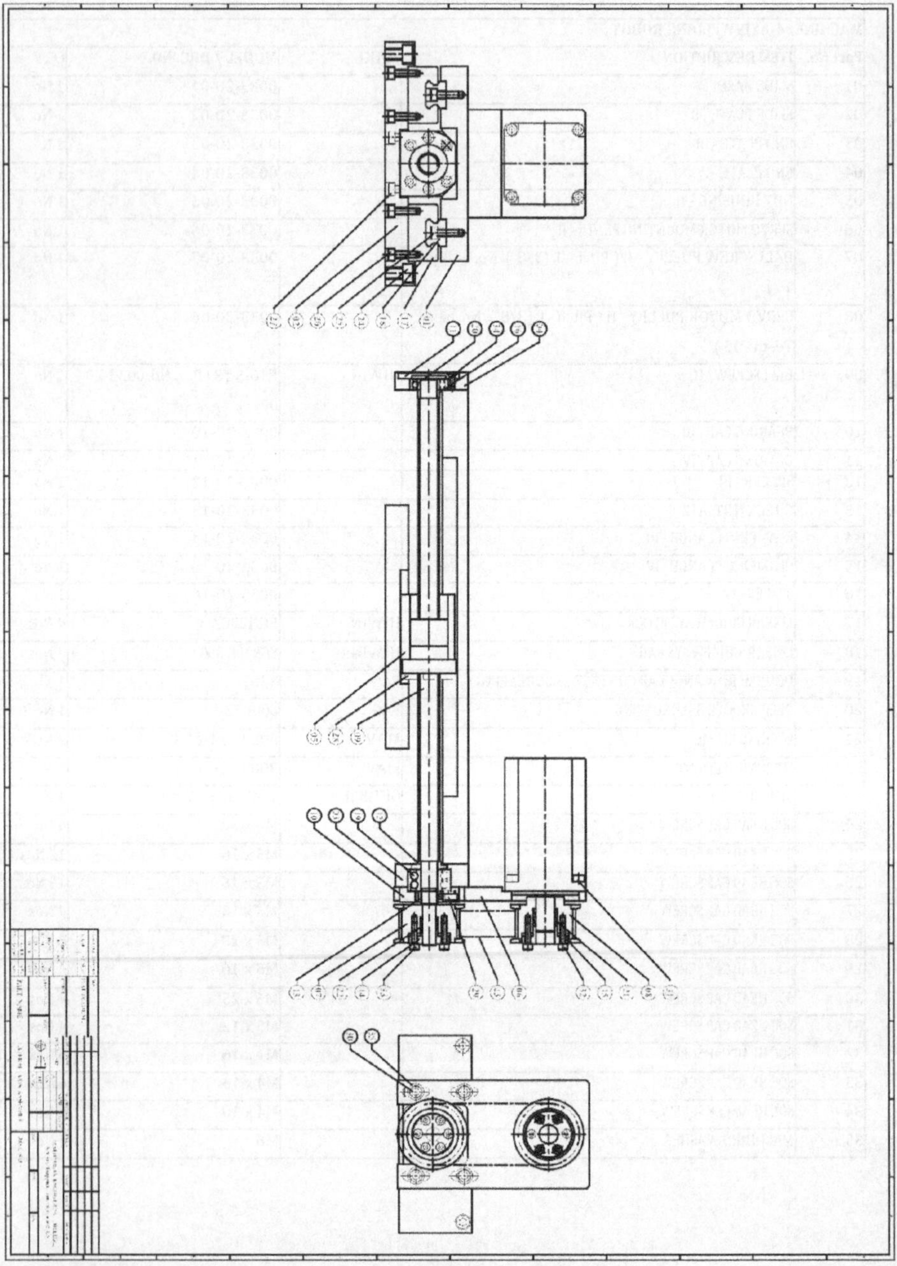

PARTS LIST				
MACHINE : 4-AXIS WELDING ROBOT				
Part No.	**ITEM DESCRIPTION**	**MAKE**	**MODEL / DRG. NO.**	**QTY**
01	SLIDE BASE - II		0033-30-01	1 No
02	SLIDE PLATE - III		0033-30-02	1 No
03	RIGHT ANGLE BRACKET		0033-30-03	1 No
04	SERVO MOTOR MOUNTING PLATE - III		0033-30-04	1 No
05	BALLSCREW - III	HIWIN	R16-5 T3 (Drg. No. 0033-	1 No
06	LM SHAFT		0033-30-06	2 Nos
07	BALLSCREW PULLEY - III (Pitch - L (3/8"), No. of Teeth - 15)		0033-30-07	1 No
08	SERVO MOTOR PULLEY - III (Pitch - L (3/8"), No. of Teeth - 15)		0033-30-08	1 No
09	BEARING CAP - V		0033-30-09	1 No
10	BEARING CAP - VI		0033-30-10	1 No
11	CHECK NUT M12		0033-30-11	1 No
12	SPACER - V		0033-30-12	1 No
13	SPACER - VI		0033-30-13	1 No
14	SPACER - VII		0033-30-14	1 No
15	CLAMPING FLANGE - V		0033-30-15	1 No
16	CLAMPING FLANGE - VI		0033-30-16	1 No
17	DOUBLE ROW ANGULAR CONTACT BALL BEARING	FAG	3201	1 No
18	DEEP GROOVE BALL BEARING	FAG	6001 ZZ	1 No
19	LOCKING ELEMENT	MAV	3003 - 14 x 18	2 Nos
20	LOCKING ELEMENT	MAV	3003 - 10 x 13	2 Nos
21	TIMING BELT	OPTIBELT	124L, 3/4" Width	1 No
22	LM BUSH BEARING	SAMICK	LME16 UU	4 Nos
23	SOC HEAD CAP SCREW		M8 x 16	4 Nos
24	SOC HEAD CAP SCREW		M8 x 25	4 Nos
25	SOC HEAD CAP SCREW		M5 x 20	6 Nos
26	SOC HEAD CAP SCREW		M5 x 16	4 Nos
27	SOC HEAD CAP SCREW		M6 x 25	4 Nos
28	SOC HEAD CAP SCREW		M4 x 16	6 Nos
29	SOC HEAD CAP SCREW		M4 x 16	6 Nos
30	SOC HEAD CAP SCREW		M5 x 12	4 Nos
31	MACHINED WASHER		M6	4 Nos
32	INTERNAL CIRCLIP		B26	4 Nos

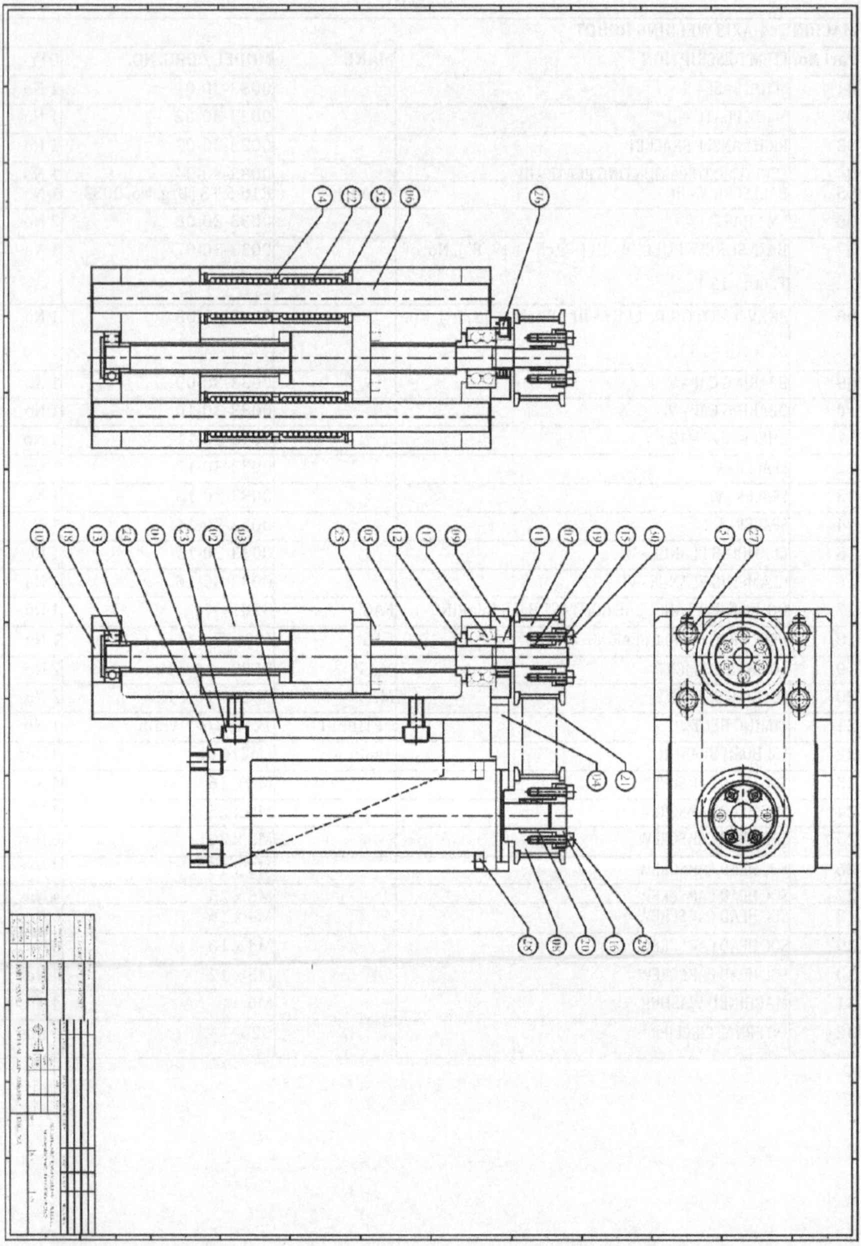

PARTS LIST				
MACHINE : 4-AXIS WELDING ROBOT				
Part No.	**ITEM DESCRIPTION**	**MAKE**	**MODEL / DRG. NO.**	**QTY**
01	GEAR BOX MOUNTING PLATE		0033-40-01	1 No
02	OUTPUT SHAFT		0033-40-02	1 No
03	SWING ARM		0033-40-03	1 No
04	STEPPER MOTOR ADAPTOR		0033-40-04	1 No
05	INPUT SLEEVE		0033-40-05	1 No
06	TORCH HOLDER		0033-40-06	1 No
07	TORCH CLAMP		0033-40-07	2 Nos
08	END CAP		0033-40-08	1 No
09	WORM GEAR BOX	TRANSTECH	MODEL - TW30 U100 F63	1 No
10	SOC HEAD CAP SCREW		M6 x 16	4 Nos
11	SOC HEAD CAP SCREW		M6 x 16	4 Nos
12	SOC HEAD CAP SCREW		M6 x 20	4 Nos
13	SOC HEAD CAP SCREW		M6 x 20	1 No
14	SOC HEAD CAP SCREW		M6 x 16	4 Nos
15	SOC HEAD CAP SCREW		M4 x 12	4 Nos

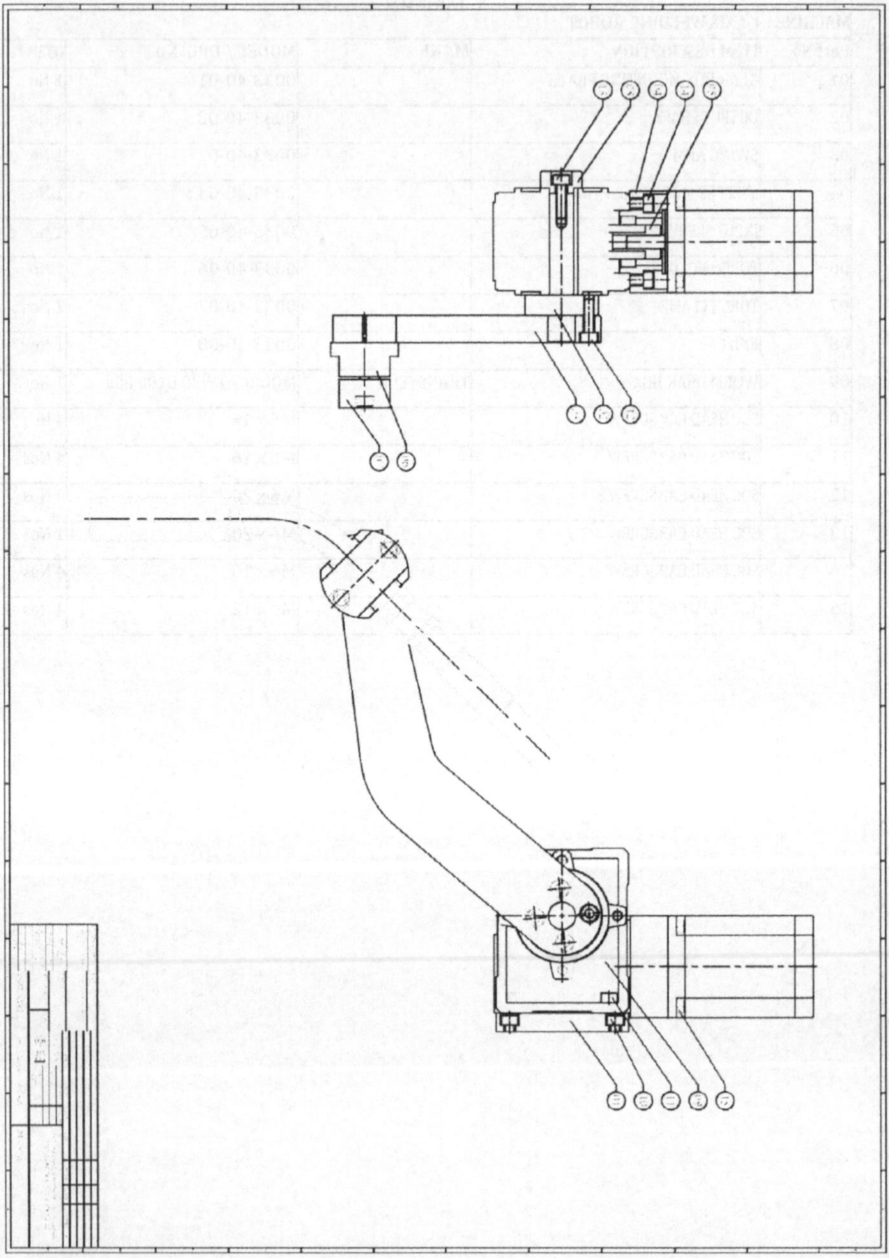

BILL OF MATERIALS - ELECTRICAL AND ELECTRONIC ITEMS				
ITEM	**DESCRIPTION**	**MAKE**	**MODEL**	**QTY**
1	CONTROL TRANSFORMER	INDIAN TRANSFORMER	TRX-500VA	1
2	SMPS POWER SUPPLY 24V DC, 5A	MEANWELL	DR-120-24	1
3	MCB 2P 10A	SCHNEIDER	XC60 C10A	3
4	MCB 2P 6A	SCHNEIDER	XC60 C6A	4
5	MCB 2P 2A	SCHNEIDER	XC60 C2A	1
6	FUSE HOLDER WITH FUSE 4A	SIEMENS	3NW20 NNSF	1
7	MCCB 4P 32A	MITSUBISHI	BHW-T10-C32	1
8	PLC 32 DI, 32DO	MITSUBISHI	FX5U-64MT/ESS	1
9	8 CHANNEL RELAY BOARD	TRINITY TOUCH	TT-1MRB-08024D2S-C	1
10	4" EXHAUST FAN	ELECTRO	REC-22038A2T	2
11	4" FAN FILTER	ELECTRO	4" FILTER	1
12	STEPPER DRIVER 4.5A	BOLANATH	BS-MSD-4.5	1
13	STEPPER MOTOR 4A, 3.1Nm/31Kgcm	BOLANATH	BH60SH86-4004-AF	1
14	4 AXIS POSITION CONTROLLER	MITSUBISHI	FX5-40SSC-S	1
15	0.3 M SSCNET III CABLE	MITSUBISHI	MRJ3 BUS 03 M	3
16	1 M SSCNET III CABLE	MITSUBISHI	MRJ3 BUS 10 M	1
17	SERVO AMPLIFIER (Y AXIS)	MITSUBISHI	MR-JE-70B	1
18	SERVO AMPLIFIER (X , Z AXIS)	MITSUBISHI	MR-JE-40B	2
19	SERVO MOTOR (Y AXIS)	MITSUBISHI	HG KN73J	1
20	SERVO MOTOR (X AXIS)	MITSUBISHI	HG KN43J	1
21	SERVO MOTOR (Z AXIS)	MITSUBISHI	HG-KN43BJ	1
22	ENCODER CABLE	MITSUBISHI	MR-J3ENCBL5M-A2-L	3
23	7" TOUCH SCREEN (HMI)	MITSUBISHI	GS2107WBDA	1
24	SERVO BATTERY	MITSUBISHI	MRBAT6V1SET	2
25	BRAKE CONNECTOR	MITSUBISHI	MR-BKS2CBL03M-A2-L	1
26	POWER CONNECTOR	MITSUBISHI	MR-PWS2CBL03M-A2-L	3
27	REMOTE BOX	AWS	REM2JY	1
28	4 POSITION JOY STICK	SCHNEIDER	XD2PA24	2
29	EMERGENCY SWITCH	TEKNIC	2AML4	1
30	PUSH BUTTION	TEKNIC	2AF8/F6/F4	5
31	4 CORE MOTOR CABLE 8M	LAPP KABEL	4G X1.0	4
32	CONTROL CABLE - 18 CORE - 10M	LAPP KABEL	18G X 0.50	1
33	LIMIT SWITCH	BOHMEN	MLRLS	4
34	PROXIMITY SENSOR	OMRON	E2-M8KS04-WP-B1 2M	2

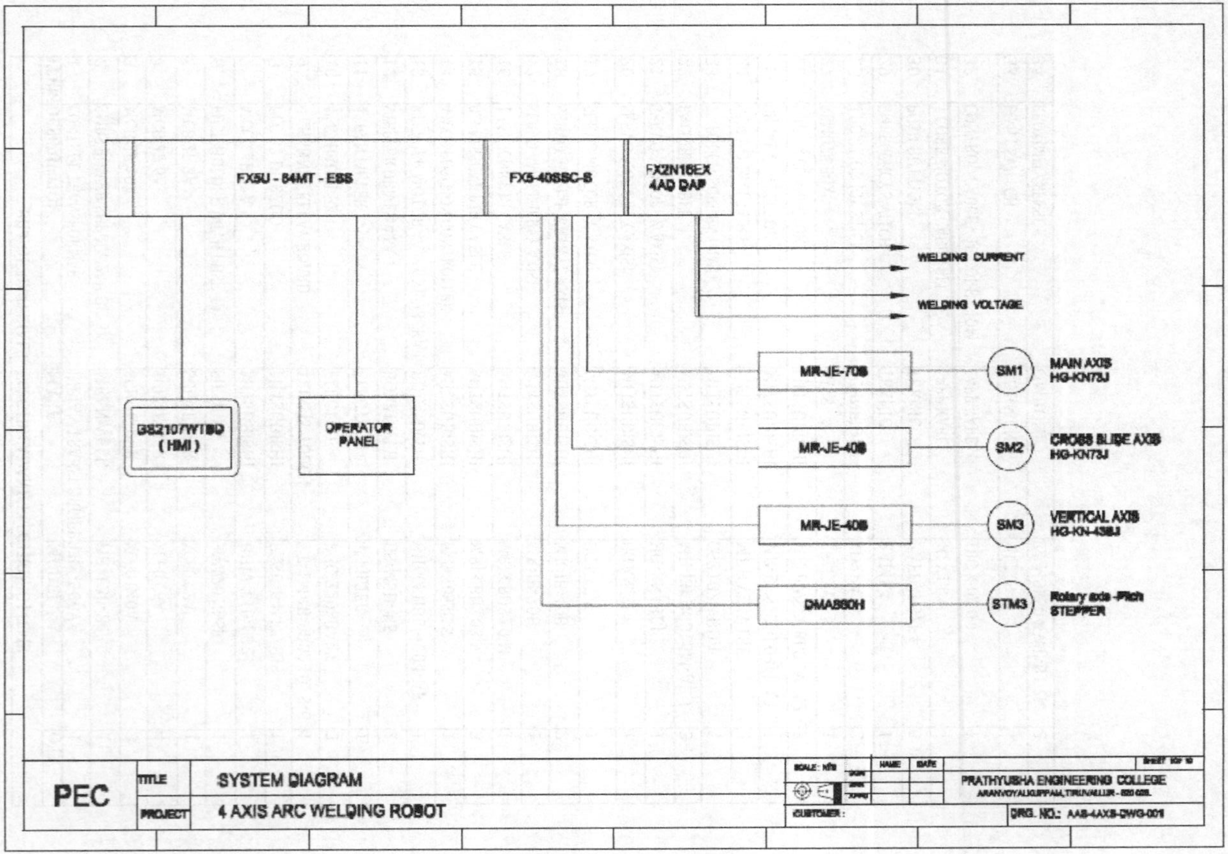
FX5U - 64MT - ESS
FX5-40SSC-S
FX2N16EX
4AD DAP
GS2107/WTBD
(HMI)
OPERATOR
PANEL
WELDING CURRENT
WELDING VOLTAGE
MR-JE-70S
MR-JE-40S
MR-JE-40S
DMA860H
SM1
MAIN AXIS
HG-KN73J
SM2
CROSS SLIDE AXIS
HG-KN73J
SM3
VERTICAL AXIS
HG-KN-43BJ
STM3
Rotary axis -Pitch
STEPPER
PEC
TITLE
SYSTEM DIAGRAM
PROJECT
4 AXIS ARC WELDING ROBOT
PRATHYUSHA ENGINEERING COLLEGE
DRG. NO.: AAB-4AXIS-DWG-001

Four Axis Arc Welding Robot

- The Four axis Arc welding Robot is designed to weld the component by using remote pendant.
- The name itself defined that this Arc Welding Robot can be moved about the four axis

Design Specification of Robot

Number of axis	:	Four
Load carrying capacity	:	50 kg
Maximum rapid	:	9000mm/min
Reach and Stroke	:	2000 mm & 120 mm
Tool orientation	:	perpendicular
Repeatability	:	± 0.2 mm
Precision and accuracy	:	± 0.2 mm
Operating environment	:	Hazardous (GMAW & FCAW)

Main Power On/Off Switch

- ON switch provides the initial power supply to the entire Arc Welding Robot System.
- OFF switch shuts the power on the entire Arc Welding Robot System.

Important Components

- PLC CONTROL SYSTEM
- ARC WELDING ROBOT
- REMOTE PENDANT

PLC Control System

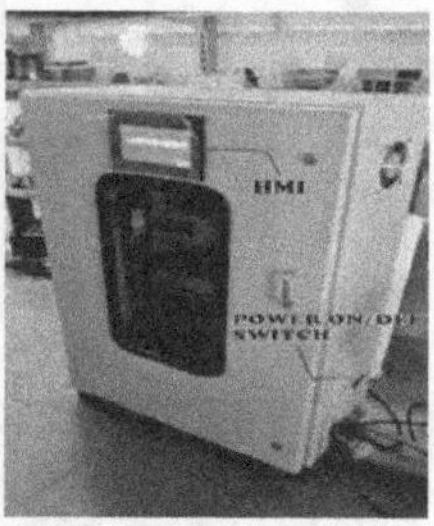

The PLC control system consists of the following components

- Driving Unit
- Programming Unit

- HMI Display Unit
- Switching Components
- Power supply unit

Driving Unit

- It consists of Servo Amplifier, Servo Motors and Stepper Drive which are having the specifications as shown in the table.

Axis	Model/Make		Capacity
Axis-1	Servo Amplifier	MR-JE-70B	2.4 Nm, 750 W, 3000 rpm, 4.8 A
	Servo Motor	HG KR73 B	
Axis-2	Servo Amplifier	MR-JE-40B	1.3 Nm, 400 W, 3000 rpm, 2.6 A
	Servo Motor	HG KR73 B	
Axis-3	Servo Amplifier	MR-JE-40B	1.3 Nm, 400 W, 3000 rpm, 2.6 A
	Servo Motor	HG KR73 B	
Axis-4	Steppe Drive& Motor	DMA860H	3.1 Nm Holding Torque, 4.5 amps

Programming Unit

- This unit comprises of the following parts in it to control the entire system.
- PLC unit Model: FX5U-64MTES/UL 32di/32do
- Position Module with SSNET communication system

Installation Procedure

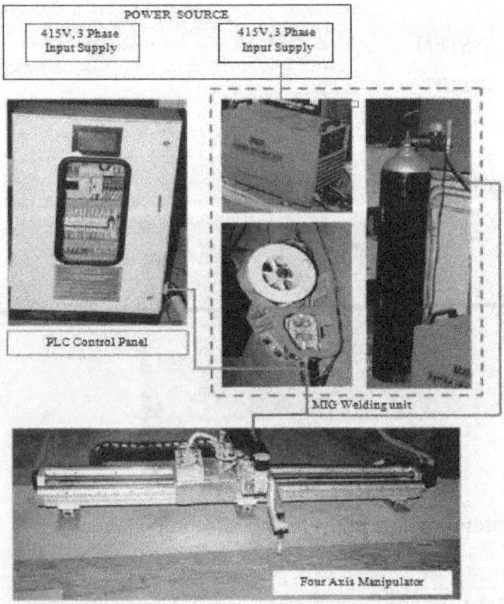

Installation of four axis arc welding robot - photographic views

Block Diagram

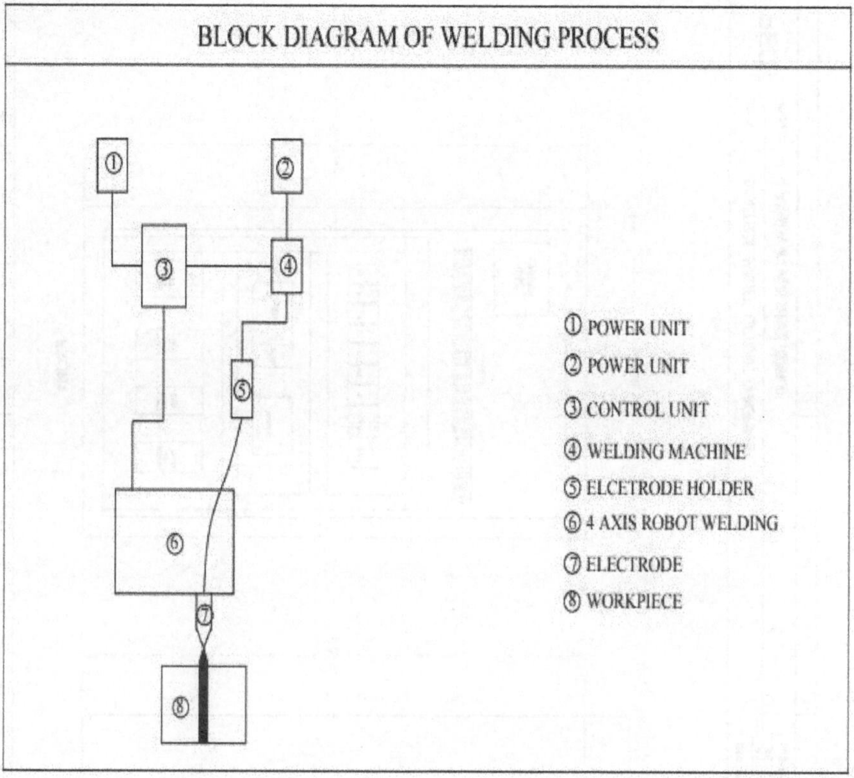

The Wiring diagram of control system are attached as Appendix 'A' as follows.

Sl	Description	Remarks
01	Control panel layout diagram	
02	Terminal wiring connection diagram	
03	PLC output wiring diagram	
04	24V Circuit wiring diagram	
05	Power circuit wiring diagram - 1	
06	Power circuit wiring diagram - 2	
07	System Diagram	

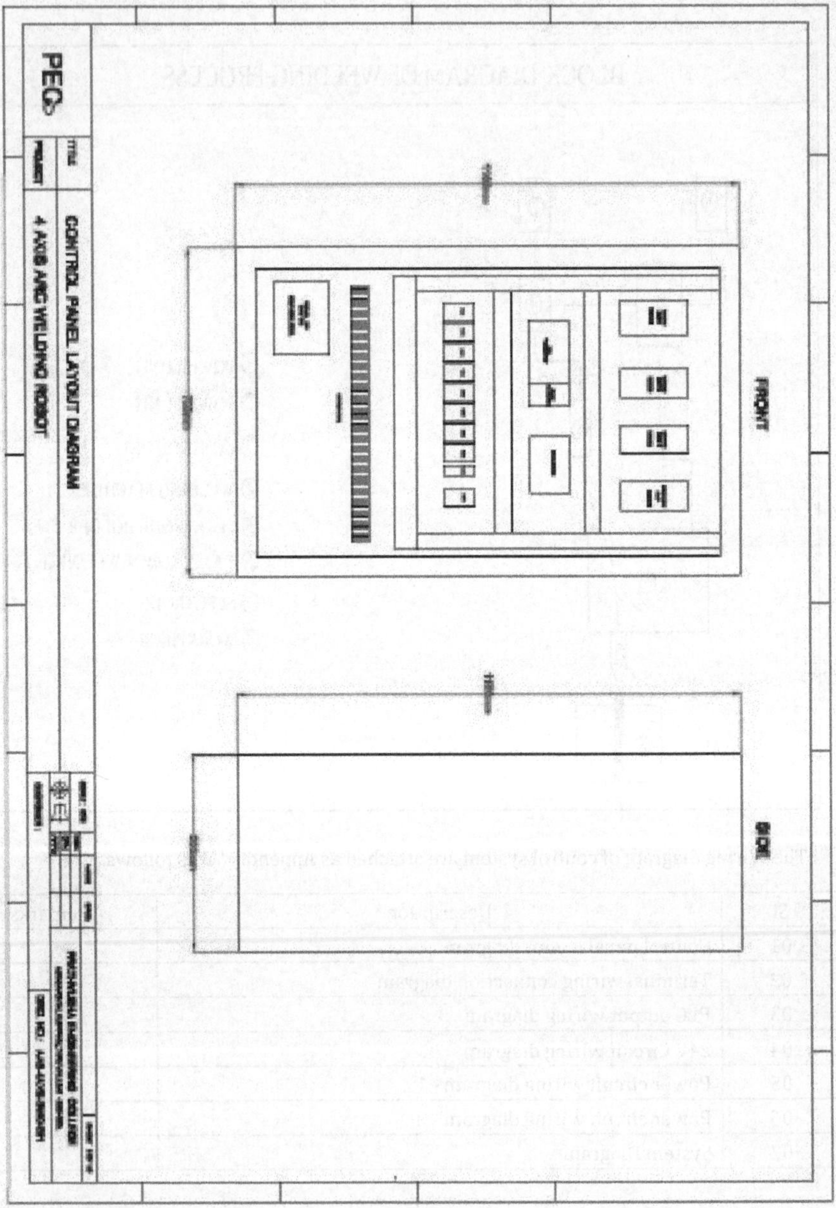
TITLE
CONTROL PANEL LAYOUT DIAGRAM
PROJECT
4 AXIS ARC WELDING ROBOT
FRONT
SIDE

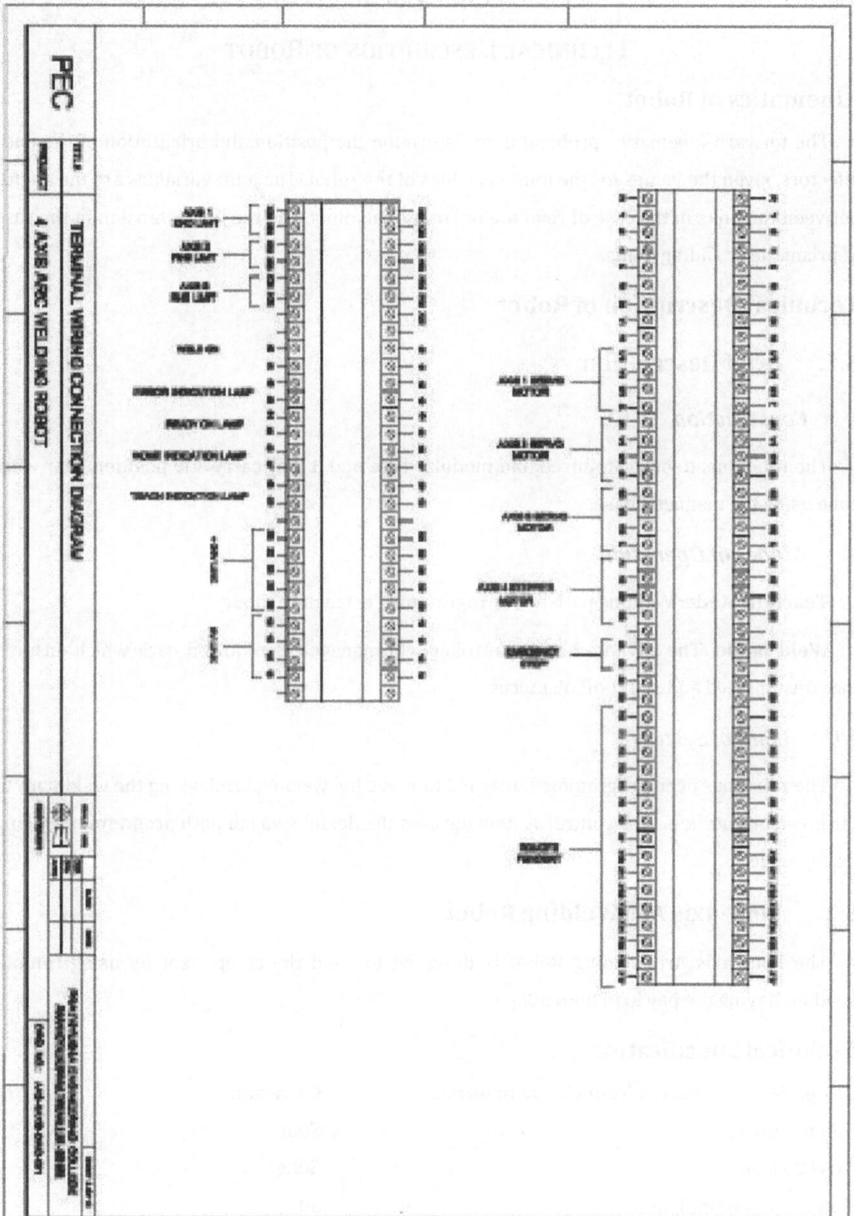
PEC
TITLE
TERMINAL WIRING CONNECTION DIAGRAM
PROJECT
4 AXIS ARC WELDING ROBOT
ERROR INDICATION LAMP
READY ON LAMP
HOME INDICATION LAMP
TEACH INDICATION LAMP
EMERGENCY STOP
REMOTE PENDANT

CHAPTER 5

TECHNICAL DESCRIPTION OF ROBOT

Kinematics of Robot

The forward kinematics problem is to determine the position and orientation of the end-effectors, given the values for the joint variables of the robot. The joint variables are the angles between the links in the case of revolute or rotational joints, and the link extension in the case of prismatic or sliding joints.

Technical Description of Robot

5.1. Brief Description

i. Construction

The robot has to be customized and modular type and it may carry and position near weld zone using the magnetic base.

ii. Modes of Operation

Teach-in Mode: Welding path to be programmed in teach-in mode

Weld mode: The advanced axis controller will manage the required path which ensures the constant feed and stand-off distances

iii. Control System

The robot has been programmed/teached to move the welding torch along the weld path to the given orientation. This control system ensures the flexible way of path programming using teach pendent.

5.2. Four-axis Arc Welding Robot

The Four axis Arc welding Robot is designed to weld the component by using remote pendant having the pay load upto 50kgs.

Technical Specification

Design Specification of Robot Type of Robot	:	**Cartesian**
Number of axis	:	Four
Axis Pay load	:	50kg
Minimum feed (mm/min)	:	30
Maximum rapid	:	9000 mm/min

Reach and Stroke	:	2000 mm & 120 mm
Tool orientation	:	perpendicular (down hand)
Repeatability	:	±0.2mm
Precision and accuracy	:	±0.2mm
Operating Environment	:	Hazardous (GMAW / FCAW)

Arc Welding Robot

Strictly the Arc Welding Robot has to be fixed at extremely close to the weld joint.

5.3. Axes of Robot

- Longitudinal Axis
- Cross slide Axis
- Vertical Axis
- Rotational Axis (PITCH)

5.4. Specifications

Axis	Stroke Length (Max) mm	Travel Speed	
		Min	Max
Longitudinal Axis	2000	30	9000
Cross Slide Axis	120	30	4500
Vertical Axis	70	30	3000
Rotational Axis	180°	30	1200

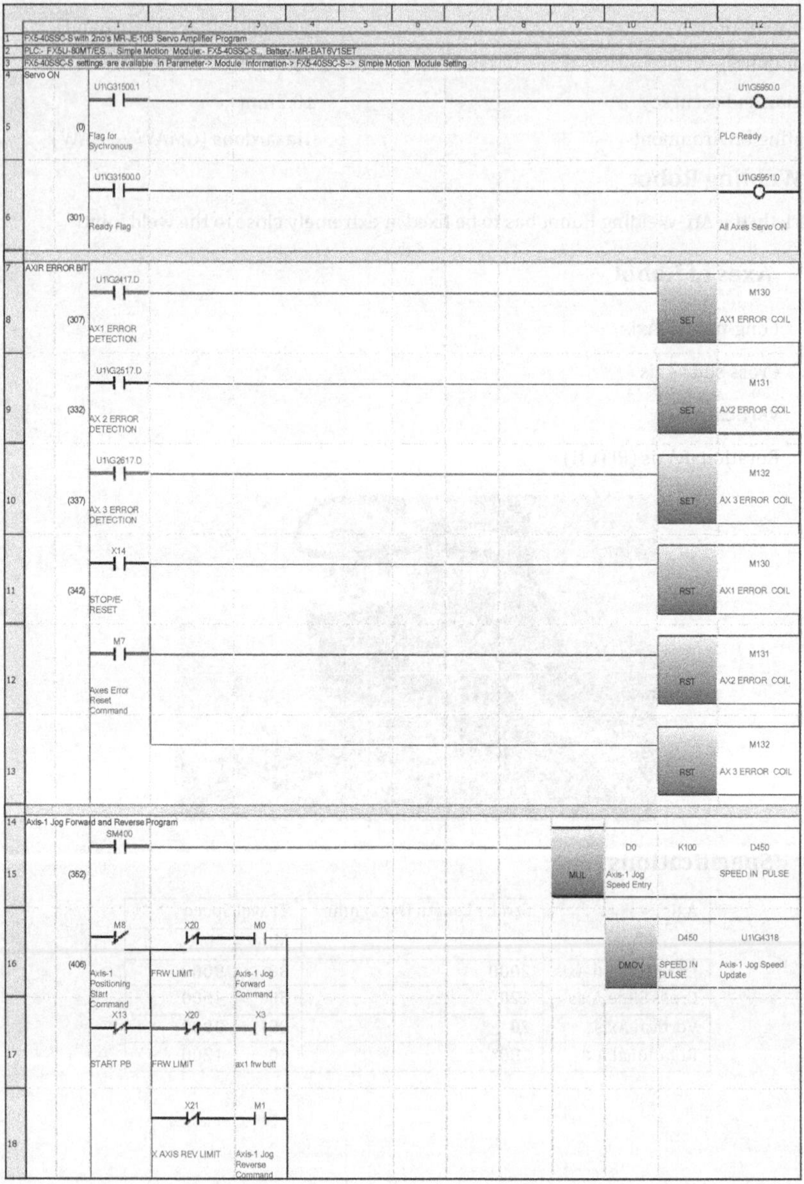
FX5-40SSC-S with 2no's MR-JE-10B Servo Amplifier Program
PLC:- FX5U-80MT/ES., Simple Motion Module:- FX5-40SSC-S., Battery:-MR-BAT6V1SET
FX5-40SSC-S settings are available in Parameter-> Module Information-> FX5-40SSC-S-> Simple Motion Module Setting
Servo ON
U1\G31500.1
(0)
Flag for Sychronous
U1\G5950.0
PLC Ready
U1\G31500.0
(301)
Ready Flag
U1\G5951.0
All Axes Servo ON
AXIR ERROR BIT
U1\G2417.D
(307)
AX1 ERROR DETECTION
SET
M130
AX1 ERROR COIL
U1\G2517.D
(332)
AX 2 ERROR DETECTION
SET
M131
AX2 ERROR COIL
U1\G2617.D
(337)
AX 3 ERROR DETECTION
SET
M132
AX 3 ERROR COIL
X14
(342)
STOP/E-RESET
RST
M130
AX1 ERROR COIL
M7
Axes Error Reset Command
RST
M131
AX2 ERROR COIL
RST
M132
AX 3 ERROR COIL
Axis-1 Jog Forward and Reverse Program
SM400
(352)
MUL
D0
Axis-1 Jog Speed Entry
K100
D450
SPEED IN PULSE
M8
(406)
Axis-1 Positioning Start Command
X20
FRW LIMIT
M0
Axis-1 Jog Forward Command
DMOV
D450
SPEED IN PULSE
U1\G4318
Axis-1 Jog Speed Update
X13
START PB
X20
FRW LIMIT
X3
ax1 frw butt
X21
X AXIS REV LIMIT
M1
Axis-1 Jog Reverse Command

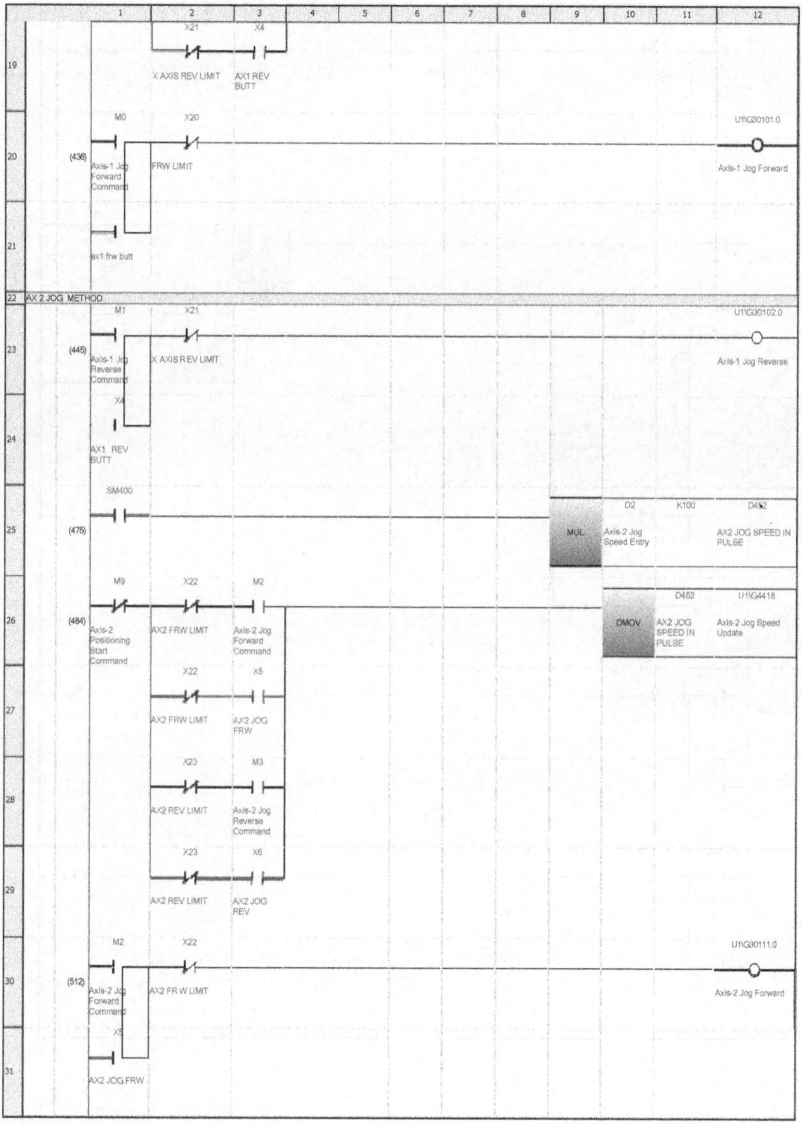
X21
X4
X AXIS REV LIMIT
AX1 REV BUTT
M0
X20
U1\G30101.0
(436)
Axis-1 Jog Forward Command
FRW LIMIT
Axis-1 Jog Forward
ax1 frw butt
AX 2 JOG METHOD
M1
X21
U1\G30102.0
(445)
Axis-1 Jog Reverse Command
X AXIS REV LIMIT
Axis-1 Jog Reverse
X4
AX1 REV BUTT
SM400
(475)
MUL
D2
K100
D452
Axis-2 Jog Speed Entry
AX2 JOG SPEED IN PULSE
M9
X22
M2
(484)
Axis-2 Positioning Start Command
AX2 FRW LIMIT
Axis-2 Jog Forward Command
DMOV
D452
U1\G4418
AX2 JOG SPEED IN PULSE
Axis-2 Jog Speed Update
X22
X5
AX2 FRW LIMIT
AX2 JOG FRW
X23
M3
AX2 REV LIMIT
Axis-2 Jog Reverse Command
X23
X6
AX2 REV LIMIT
AX2 JOG REV
M2
X22
U1\G30111.0
(512)
Axis-2 Jog Forward Command
AX2 FR W LIMIT
Axis-2 Jog Forward
X5
AX2 JOG FRW

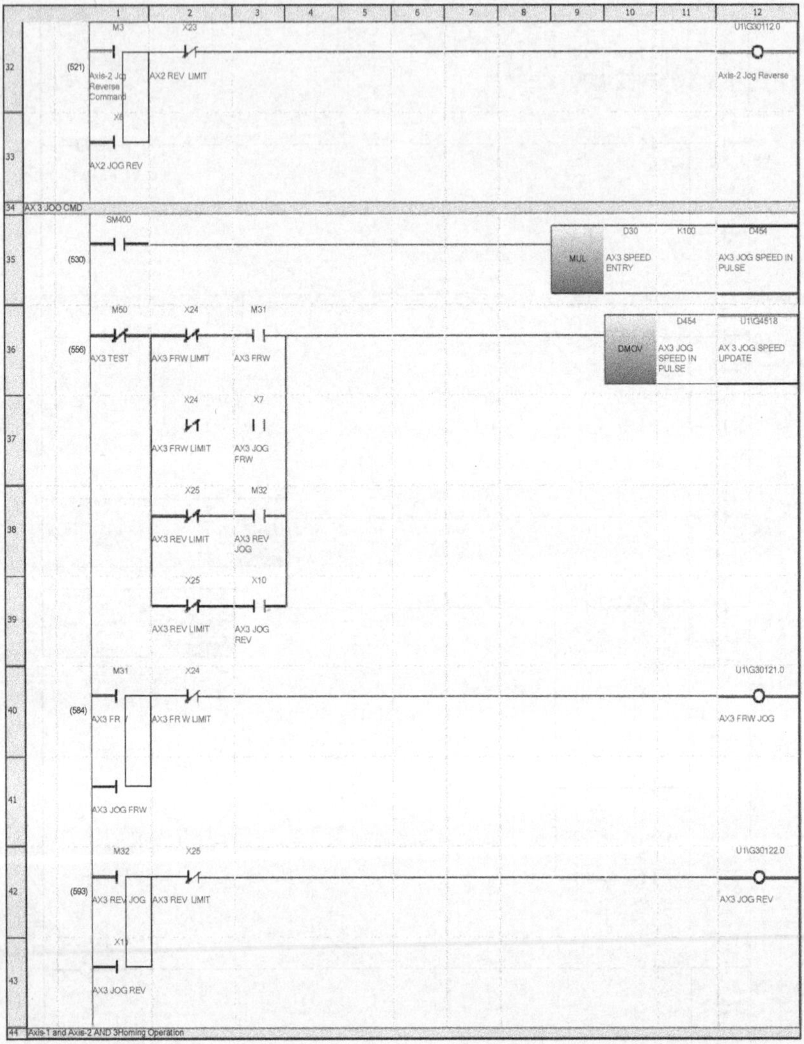
1 2 3 4 5 6 7 8 9 10 11 12
M3
X23
U1\G30112.0
32
(521)
Axis-2 Jog Reverse Command
AX2 REV LIMIT
Axis-2 Jog Reverse
33
AX2 JOG REV
34 AX 3 JOG CMD
SM400
35
(530)
MUL
D30
AX3 SPEED ENTRY
K100
D454
AX3 JOG SPEED IN PULSE
M50
X24
M31
36
(556)
AX3 TEST
AX3 FRW LIMIT
AX3 FRW
DMOV
D454
AX3 JOG SPEED IN PULSE
U1\G4518
AX 3 JOG SPEED UPDATE
X24
X7
37
AX3 FRW LIMIT
AX3 JOG FRW
X25
M32
38
AX3 REV LIMIT
AX3 REV JOG
X25
X10
39
AX3 REV LIMIT
AX3 JOG REV
M31
X24
U1\G30121.0
40
(584)
AX3 FR W LIMIT
AX3 FRW JOG
41
AX3 JOG FRW
M32
X25
U1\G30122.0
42
(593)
AX3 REV LIMIT
AX3 JOG REV
43
AX3 JOG REV
44 Axis-1 and Axis-2 AND 3Homing Operation

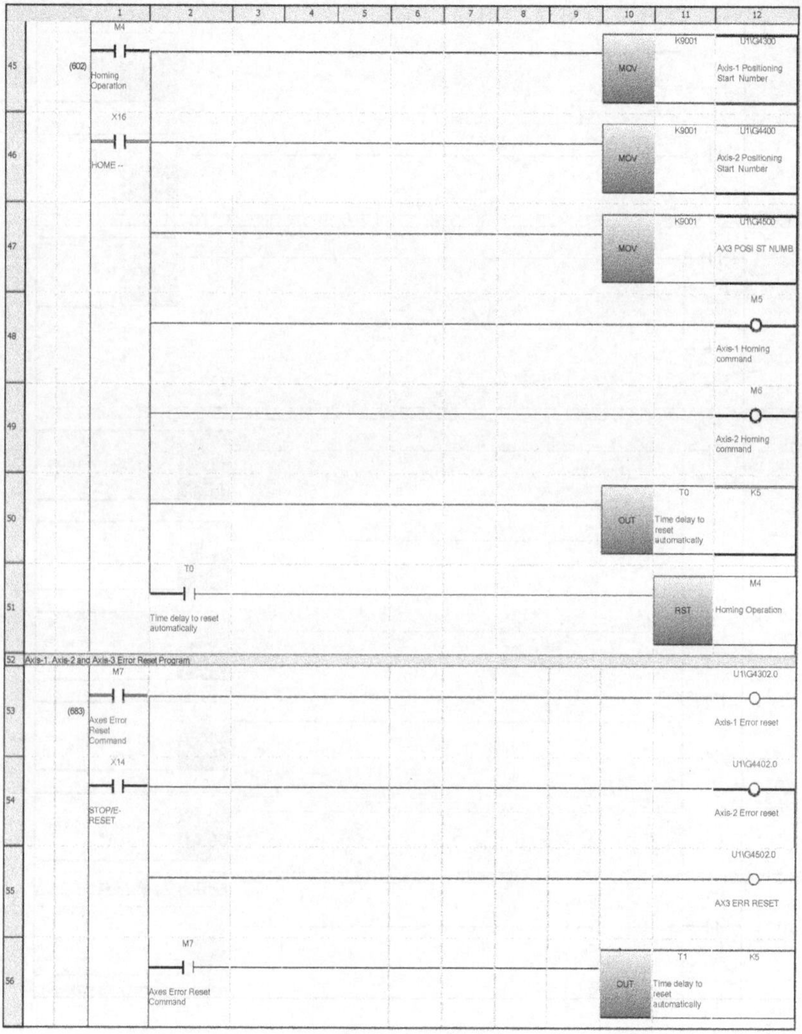
1
2
3
4
5
6
7
8
9
10
11
12
45
(602)
M4
Homing Operation
MOV
K9001
U1\G4300
Axis-1 Positioning Start Number
46
X16
HOME --
MOV
K9001
U1\G4400
Axis-2 Positioning Start Number
47
MOV
K9001
U1\G4500
AX3 POSI ST NUMB
48
M5
Axis-1 Homing command
49
M6
Axis-2 Homing command
50
OUT
T0
K5
Time delay to reset automatically
51
T0
Time delay to reset automatically
RST
M4
Homing Operation
52 Axis-1, Axis-2 and Axis-3 Error Reset Program
53
(683)
M7
Axes Error Reset Command
U1\G4302.0
Axis-1 Error reset
54
X14
STOP/E-RESET
U1\G4402.0
Axis-2 Error reset
55
U1\G4502.0
AX3 ERR RESET
56
M7
Axes Error Reset Command
OUT
T1
K5
Time delay to reset automatically

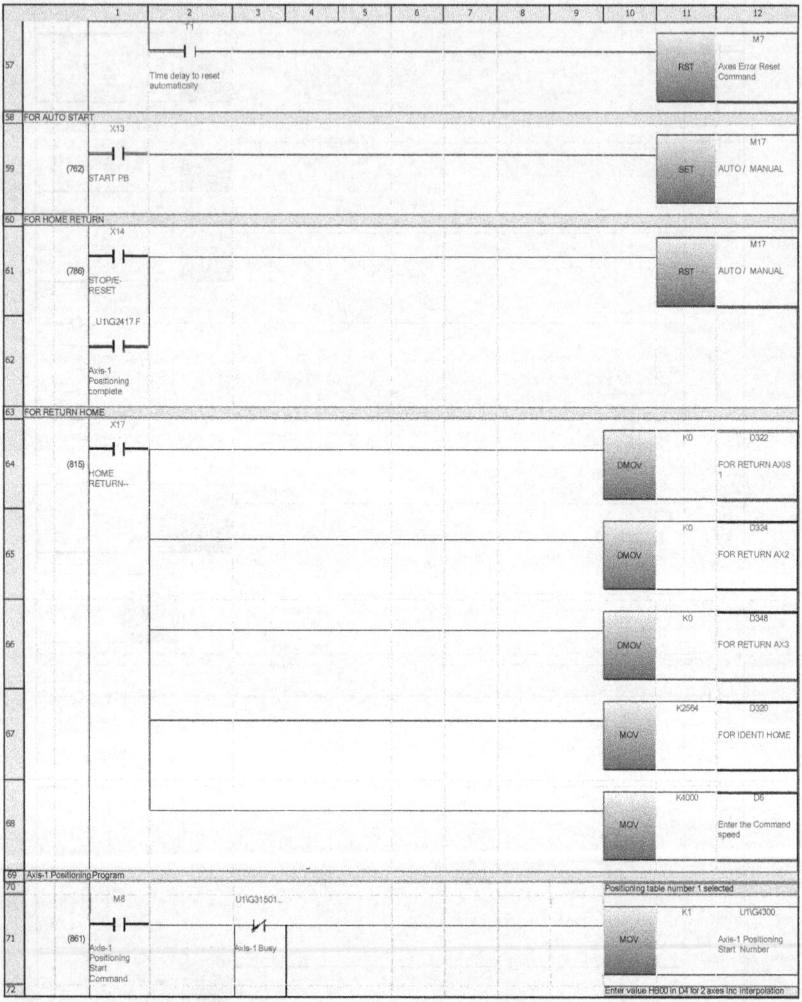
1 2 3 4 5 6 7 8 9 10 11 12
57
T1
Time delay to reset automatically
RST
M7
Axes Error Reset Command
58 FOR AUTO START
59 (762)
X13
START PB
SET
M17
AUTO / MANUAL
60 FOR HOME RETURN
61 (786)
X14
STOP/E-RESET
RST
M17
AUTO / MANUAL
62
U1\G2417.F
Axis-1 Positioning complete
63 FOR RETURN HOME
64 (815)
X17
HOME RETURN--
DMOV
K0
D322
FOR RETURN AXIS 1
65
DMOV
K0
D324
FOR RETURN AX2
66
DMOV
K0
D348
FOR RETURN AX3
67
MOV
K2564
D320
FOR IDENTI HOME
68
MOV
K4000
D6
Enter the Command speed
69 Axis-1 Positioning Program
70
Positioning table number 1 selected
71 (861)
M8
Axis-1 Positioning Start Command
U1\G31501..
Axis-1 Busy
MOV
K1
U1\G4300
Axis-1 Positioning Start Number
72
Enter value HB00 in D4 for 2 axes Inc interpolation

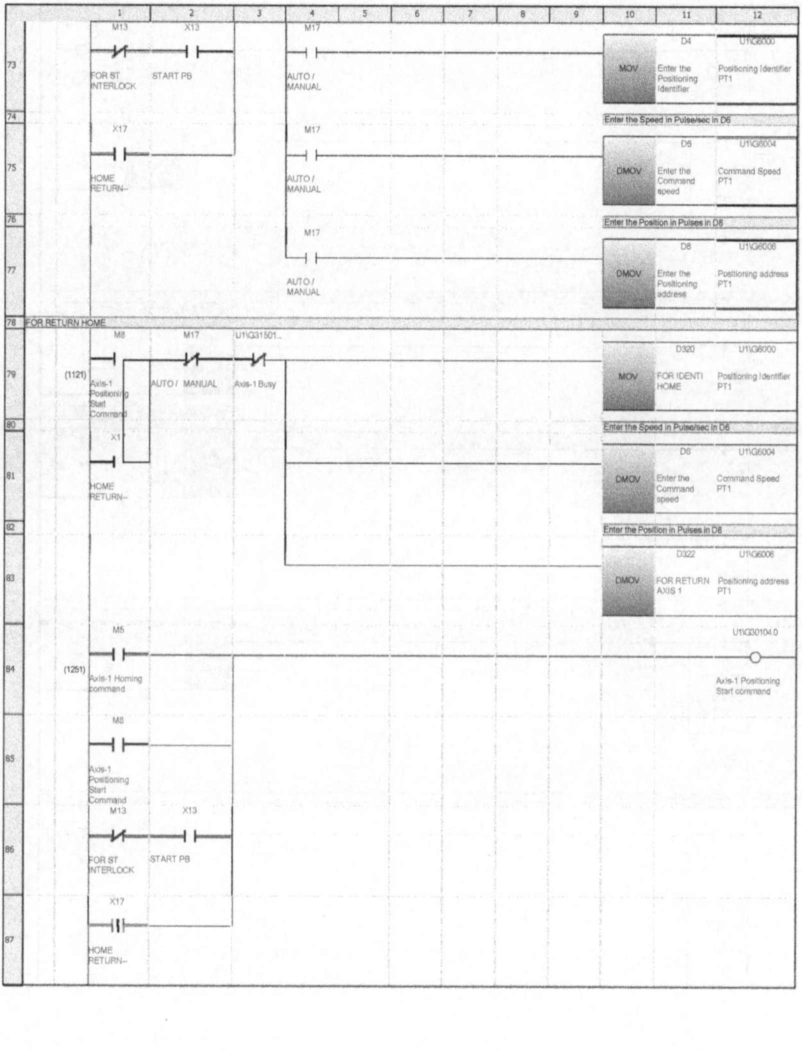
M13
X13
M17
FOR ST INTERLOCK
START PB
AUTO / MANUAL
MOV
D4
U1\G6000
Enter the Positioning Identifier
Positioning Identifier PT1
Enter the Speed in Pulse/sec in D6
X17
HOME RETURN--
DMOV
D6
U1\G6004
Enter the Command speed
Command Speed PT1
Enter the Position in Pulses in D8
D8
U1\G6006
Enter the Positioning address
Positioning address PT1
FOR RETURN HOME
M8
U1\G31501...
(1121)
Axis-1 Positioning Start Command
Axis-1 Busy
D320
FOR IDENTI HOME
D322
FOR RETURN AXIS 1
M5
(1251)
Axis-1 Homing command
U1\G30104.0
Axis-1 Positioning Start command

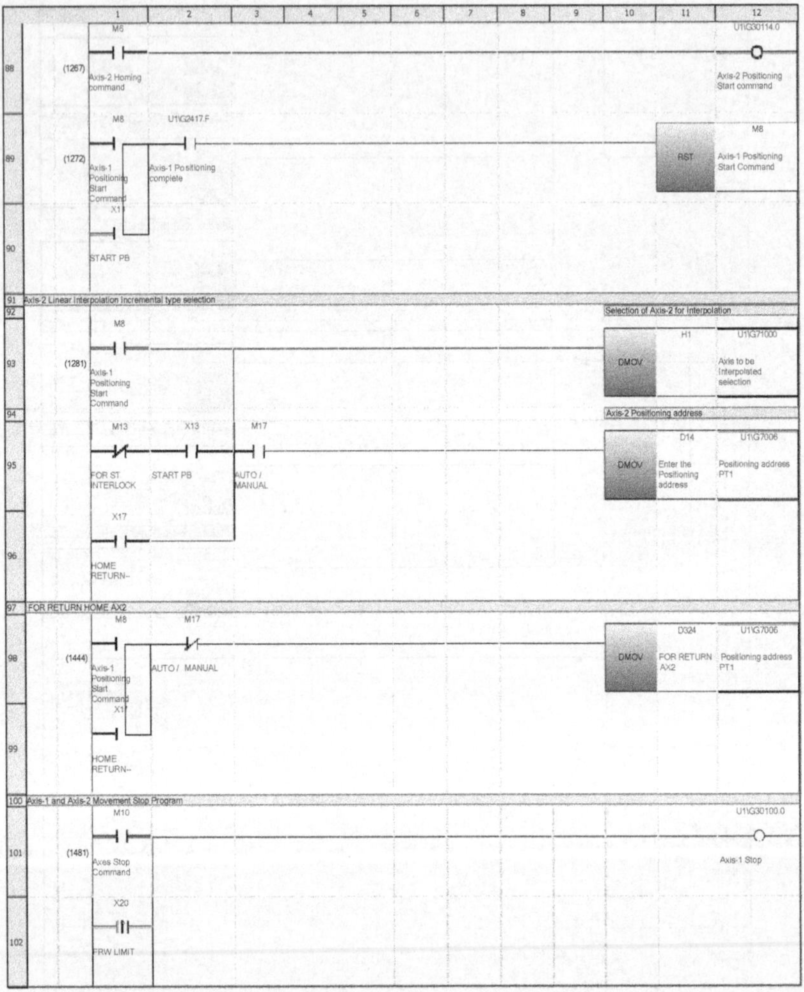
M8
U1\G30114.0
(1267)
Axis-2 Homing command
Axis-2 Positioning Start command
M8
U1\G2417.F
(1272)
Axis-1 Positioning Start Command
Axis-1 Positioning complete
RST
M8
Axis-1 Positioning Start Command
X1
START PB
Axis-2 Linear Interpolation Incremental type selection
Selection of Axis-2 for Interpolation
M8
(1281)
Axis-1 Positioning Start Command
DMOV
H1
U1\G71000
Axis to be Interpolated selection
Axis-2 Positioning address
M13
X13
M17
FOR ST INTERLOCK
START PB
AUTO / MANUAL
DMOV
D14
U1\G7006
Enter the Positioning address
Positioning address PT1
X17
HOME RETURN--
FOR RETURN HOME AX2
M8
M17
(1444)
Axis-1 Positioning Start Command
AUTO / MANUAL
DMOV
D324
U1\G7006
FOR RETURN AX2
Positioning address PT1
X1
HOME RETURN--
Axis-1 and Axis-2 Movement Stop Program
M10
U1\G30100.0
(1481)
Axes Stop Command
Axis-1 Stop
X20
FRW LIMIT

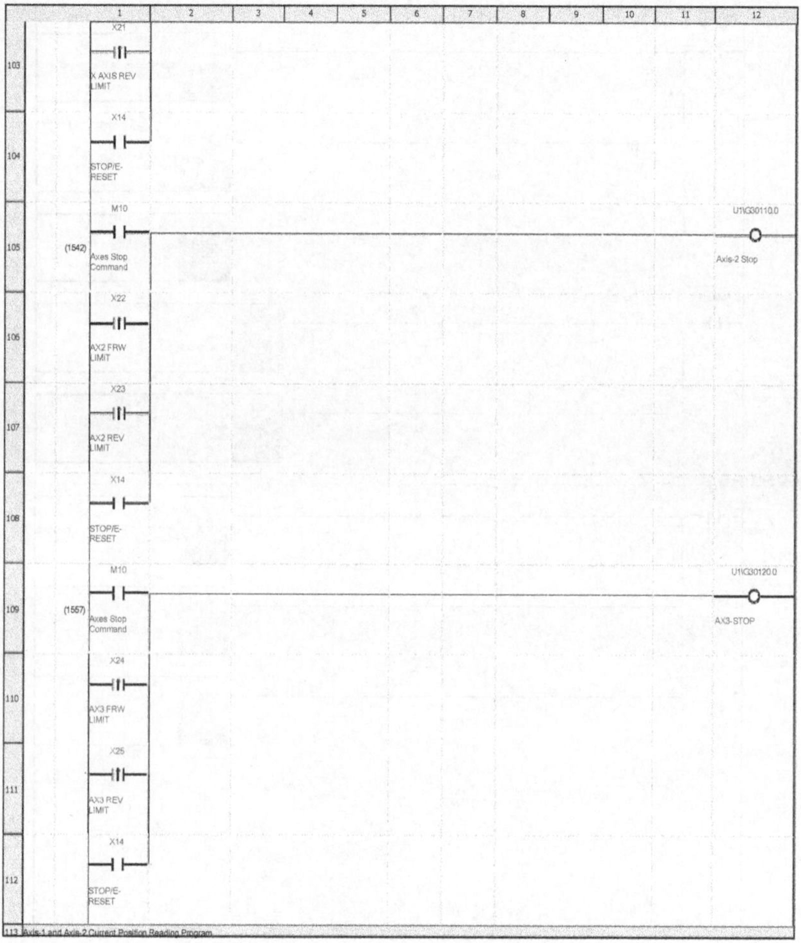
103
X21
X AXIS REV LIMIT
104
X14
STOP/E-RESET
105
(1542)
M10
Axes Stop Command
U1\G30110.0
Axis-2 Stop
106
X22
AX2 FRW LIMIT
107
X23
AX2 REV LIMIT
108
X14
STOP/E-RESET
109
(1557)
M10
Axes Stop Command
U1\G30120.0
AX3-STOP
110
X24
AX3 FRW LIMIT
111
X25
AX3 REV LIMIT
112
X14
STOP/E-RESET
113 Axis-1 and Axis-2 Current Position Reading Program

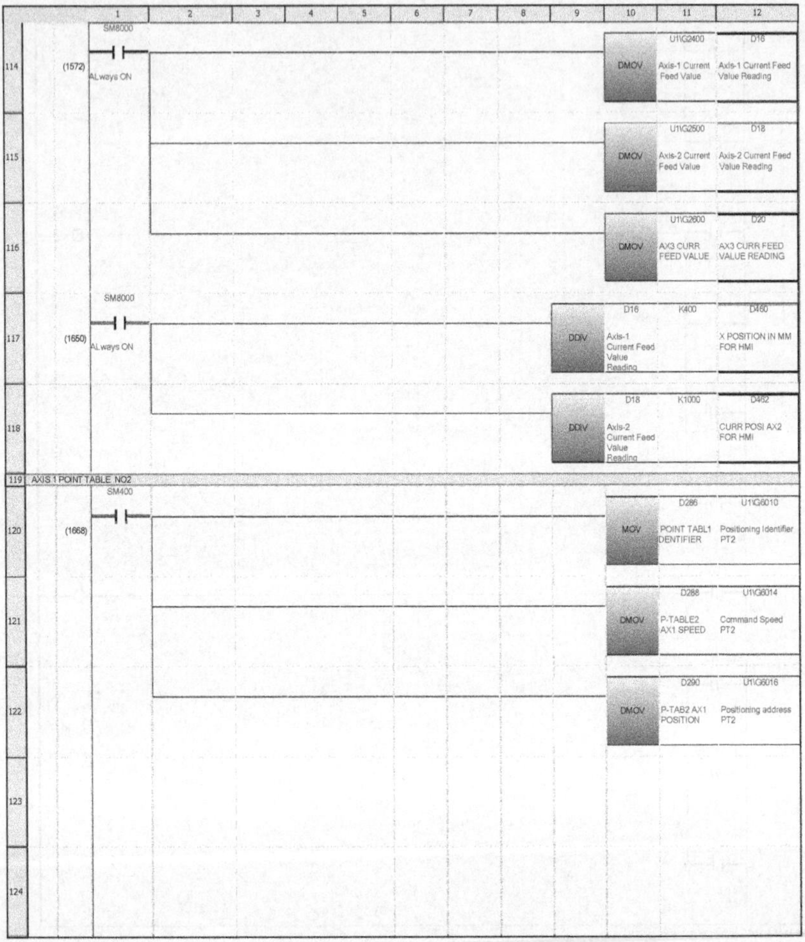
114 (1572) SM8000 ALways ON DMOV U1\G2400 Axis-1 Current Feed Value D16 Axis-1 Current Feed Value Reading
115 DMOV U1\G2500 Axis-2 Current Feed Value D18 Axis-2 Current Feed Value Reading
116 DMOV U1\G2600 AX3 CURR FEED VALUE D20 AX3 CURR FEED VALUE READING
117 (1650) SM8000 ALways ON DDIV D16 Axis-1 Current Feed Value Reading K400 D460 X POSITION IN MM FOR HMI
118 DDIV D18 Axis-2 Current Feed Value Reading K1000 D462 CURR POSI AX2 FOR HMI
119 AXIS 1 POINT TABLE NO2
120 (1668) SM400 MOV D286 POINT TABL1 DENTIFIER U1\G6010 Positioning Identifier PT2
121 DMOV D288 P-TABLE2 AX1 SPEED U1\G6014 Command Speed PT2
122 DMOV D290 P-TAB2 AX1 POSITION U1\G6016 Positioning address PT2
123
124

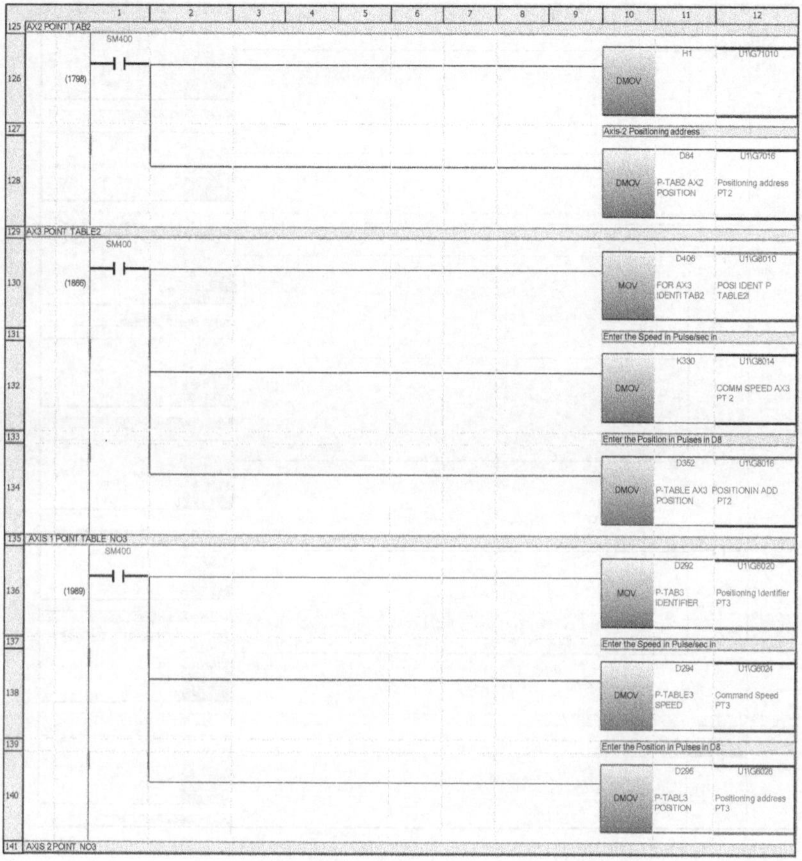
125 AX2 POINT TAB2
SM400
126 (1798)
DMOV H1 U1\G71010
127 Axis-2 Positioning address
128
DMOV D84 U1\G7016
P-TAB2 AX2 POSITION
Positioning address PT2
129 AX3 POINT TABLE2
SM400
130 (1866)
MOV D406 U1\G8010
FOR AX3 IDENTI TAB2
POSI IDENT P TABLE2I
131 Enter the Speed in Pulse/sec in
132
DMOV K330 U1\G8014
COMM SPEED AX3 PT 2
133 Enter the Position in Pulses in D8
134
DMOV D352 U1\G8016
P-TABLE AX3 POSITION
POSITIONIN ADD PT2
135 AXIS 1 POINT TABLE NO3
SM400
136 (1989)
MOV D292 U1\G6020
P-TAB3 IDENTIFIER
Positioning Identifier PT3
137 Enter the Speed in Pulse/sec in
138
DMOV D294 U1\G6024
P-TABLE3 SPEED
Command Speed PT3
139 Enter the Position in Pulses in D8
140
DMOV D296 U1\G6026
P-TABL3 POSITION
Positioning address PT3
141 AXIS 2 POINT NO3

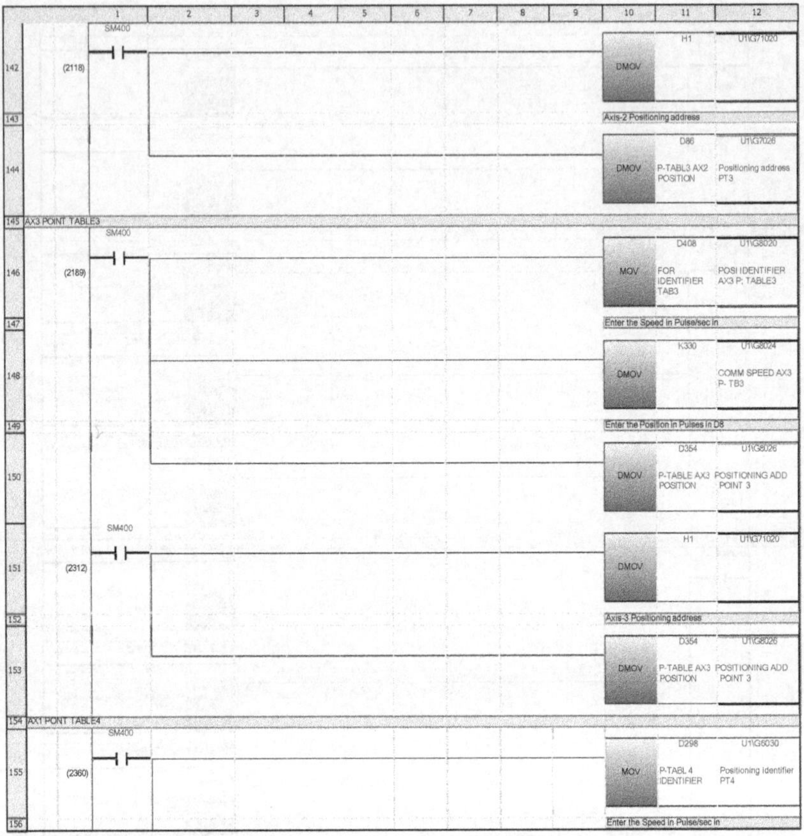
SM400
142 (2118)
DMOV H1 U1\G71020
143 Axis-2 Positioning address
144 DMOV D86 U1\G7026
P-TABL3 AX2 POSITION
Positioning address PT3
145 AX3 POINT TABLE3
SM400
146 (2189)
MOV D408 U1\G8020
FOR IDENTIFIER TAB3
POSI IDENTIFIER AX3 P; TABLE3
147 Enter the Speed in Pulse/sec in
148 DMOV K330 U1\G8024
COMM SPEED AX3 P- TB3
149 Enter the Position in Pulses in D8
150 DMOV D354 U1\G8026
P-TABLE AX3 POSITION
POSITIONING ADD POINT 3
SM400
151 (2312)
DMOV H1 U1\G71020
152 Axis-3 Positioning address
153 DMOV D354 U1\G8026
P-TABLE AX3 POSITION
POSITIONING ADD POINT 3
154 AX1 PONT TABLE4
SM400
155 (2360)
MOV D298 U1\G6030
P-TABL 4 IDENTIFIER
Positioning Identifier PT4
156 Enter the Speed in Pulse/sec in

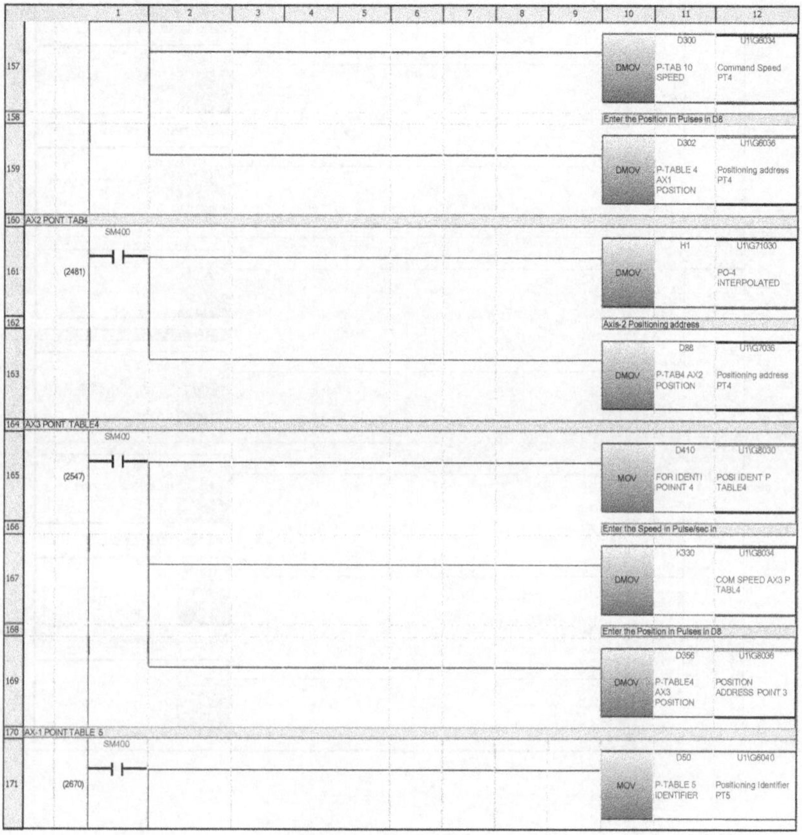
1 2 3 4 5 6 7 8 9 10 11 12
157
DMOV D300 P-TAB 10 SPEED
U1\G8034 Command Speed PT4
158 Enter the Position in Pulses in D8
159
DMOV D302 P-TABLE 4 AX1 POSITION
U1\G8036 Positioning address PT4
160 AX2 PONT TAB4
161 SM400 (2481)
DMOV H1
U1\G71030 PO-4 INTERPOLATED
162 Axis-2 Positioning address
163
DMOV D88 P-TAB4 AX2 POSITION
U1\G7036 Positioning address PT4
164 AX3 POINT TABLE4
165 SM400 (2547)
MOV D410 FOR IDENTI POINNT 4
U1\G8030 POSI IDENT P TABLE4
166 Enter the Speed in Pulse/sec in
167
DMOV K330
U1\G8034 COM SPEED AX3 P TABL4
168 Enter the Position in Pulses in D8
169
DMOV D356 P-TABLE4 AX3 POSITION
U1\G8036 POSITION ADDRESS POINT 3
170 AX-1 POINT TABLE 5
171 SM400 (2670)
MOV D50 P-TABLE 5 IDENTIFIER
U1\G6040 Positioning Identifier PT5

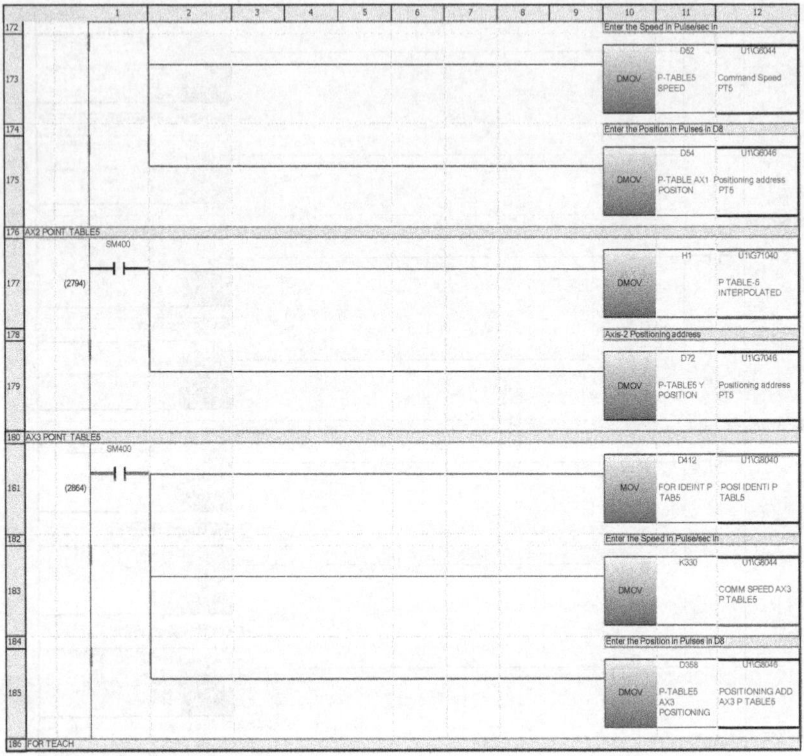
172 Enter the Speed in Pulse/sec in
173 DMOV D52 U1\G8044
P-TABLE5 SPEED
Command Speed PT5
174 Enter the Position in Pulses in D8
175 DMOV D54 U1\G8046
P-TABLE AX1 POSITON
Positioning address PT5
176 AX2 POINT TABLE5
SM400
177 (2794)
DMOV H1 U1\G71040
P TABLE-5 INTERPOLATED
178 Axis-2 Positioning address
179 DMOV D72 U1\G7046
P-TABLE5 Y POSITION
Positioning address PT5
180 AX3 POINT TABLE5
SM400
181 (2864)
MOV D412 U1\G8040
FOR IDEINT P TAB5
POSI IDENTI P TABL5
182 Enter the Speed in Pulse/sec in
183 DMOV K330 U1\G8044
COMM SPEED AX3 P TABLE5
184 Enter the Position in Pulses in D8
185 DMOV D358 U1\G8046
P-TABLE5 AX3 POSITIONING
POSITIONING ADD AX3 P TABLE5
186 FOR TEACH

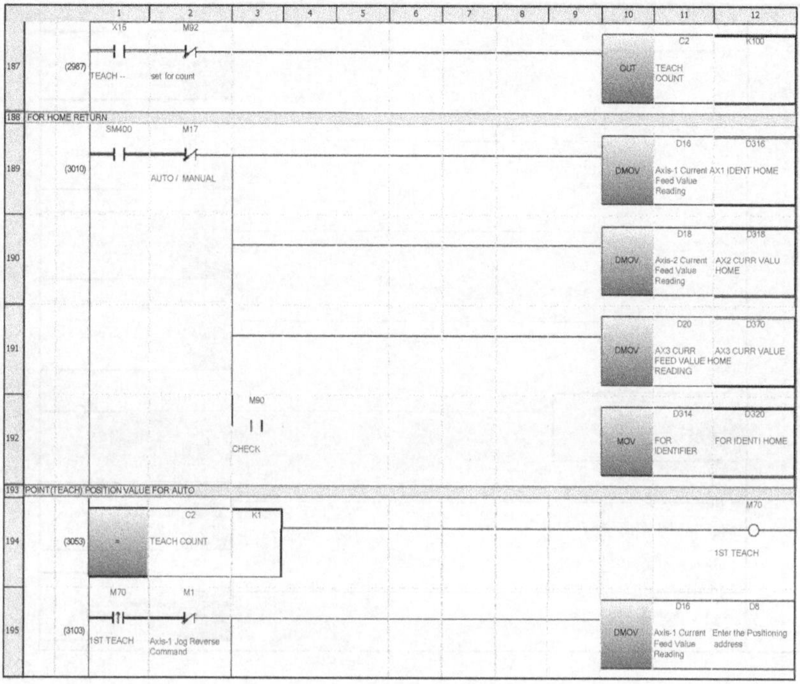
1
2
3
4
5
6
7
8
9
10
11
12
187
(2987)
X15
M92
TEACH --
set for count
OUT
C2
K100
TEACH
COUNT
188 FOR HOME RETURN
189
(3010)
SM400
M17
AUTO / MANUAL
DMOV
D16
D316
Axis-1 Current Feed Value Reading
AX1 IDENT HOME
190
DMOV
D18
D318
Axis-2 Current Feed Value Reading
AX2 CURR VALU HOME
191
DMOV
D20
D370
AX3 CURR FEED VALUE READING
AX3 CURR VALUE HOME
192
M90
CHECK
MOV
D314
D320
FOR IDENTIFIER
FOR IDENTI HOME
193 POINT(TEACH) POSITION VALUE FOR AUTO
194
(3053)
=
C2
K1
TEACH COUNT
M70
1ST TEACH
195
(3103)
M70
M1
1ST TEACH
Axis-1 Jog Reverse Command
DMOV
D16
D8
Axis-1 Current Feed Value Reading
Enter the Positioning address

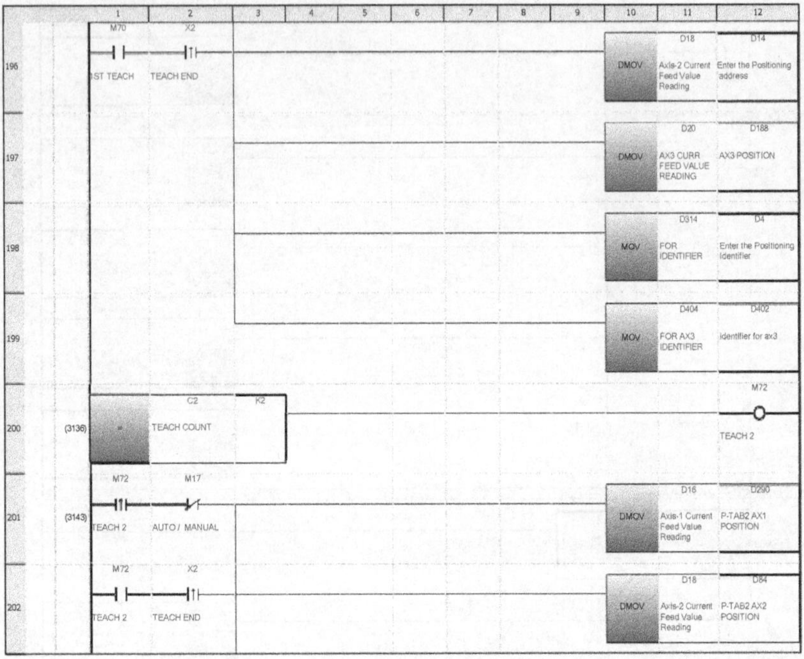
1 2 3 4 5 6 7 8 9 10 11 12
196
M70
1ST TEACH
X2
TEACH END
DMOV
D18
Axis-2 Current Feed Value Reading
D14
Enter the Positioning address
197
DMOV
D20
AX3 CURR FEED VALUE READING
D188
AX3 POSITION
198
MOV
D314
FOR IDENTIFIER
D4
Enter the Positioning Identifier
199
MOV
D404
FOR AX3 IDENTIFIER
D402
identifier for ax3
200
(3136)
=
C2
TEACH COUNT
K2
M72
TEACH 2
201
(3143)
M72
TEACH 2
M17
AUTO / MANUAL
DMOV
D16
Axis-1 Current Feed Value Reading
D290
P-TAB2 AX1 POSITION
202
M72
TEACH 2
X2
TEACH END
DMOV
D18
Axis-2 Current Feed Value Reading
D84
P-TAB2 AX2 POSITION

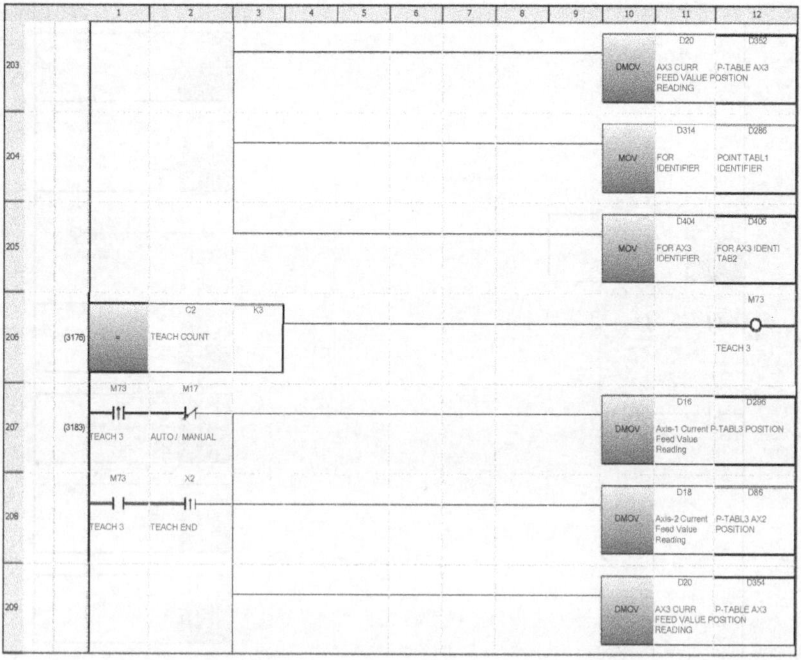
203
DMOV
D20
AX3 CURR FEED VALUE READING
D352
P-TABLE AX3 POSITION
204
MOV
D314
FOR IDENTIFIER
D286
POINT TABL1 IDENTIFIER
205
MOV
D404
FOR AX3 IDENTIFIER
D406
FOR AX3 IDENTI TAB2
206
(3176)
=
C2
TEACH COUNT
K3
M73
TEACH 3
207
(3183)
M73
TEACH 3
M17
AUTO / MANUAL
DMOV
D16
Axis-1 Current Feed Value Reading
D296
P-TABL3 POSITION
208
M73
TEACH 3
X2
TEACH END
DMOV
D18
Axis-2 Current Feed Value Reading
D86
P-TABL3 AX2 POSITION
209
DMOV
D20
AX3 CURR FEED VALUE READING
D354
P-TABLE AX3 POSITION

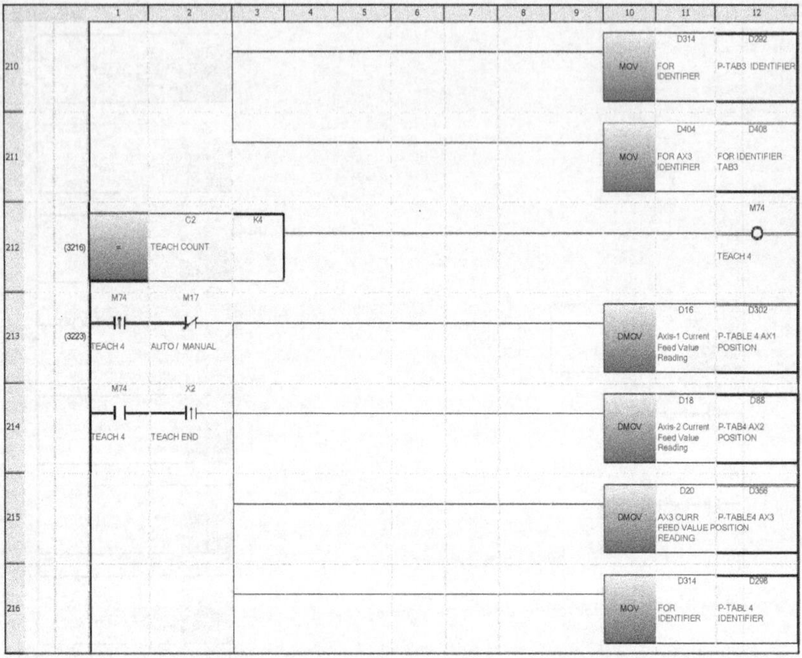
210
MOV
D314
FOR IDENTIFIER
D292
P-TAB3 IDENTIFIER
211
MOV
D404
FOR AX3 IDENTIFIER
D408
FOR IDENTIFIER TAB3
212
(3216)
=
C2
TEACH COUNT
K4
M74
TEACH 4
213
(3223)
M74
TEACH 4
M17
AUTO / MANUAL
DMOV
D16
Axis-1 Current Feed Value Reading
D302
P-TABLE 4 AX1 POSITION
214
M74
TEACH 4
X2
TEACH END
DMOV
D18
Axis-2 Current Feed Value Reading
D88
P-TAB4 AX2 POSITION
215
DMOV
D20
AX3 CURR FEED VALUE READING
D366
P-TABLE4 AX3 POSITION
216
MOV
D314
FOR IDENTIFIER
D298
P-TABL 4 IDENTIFIER

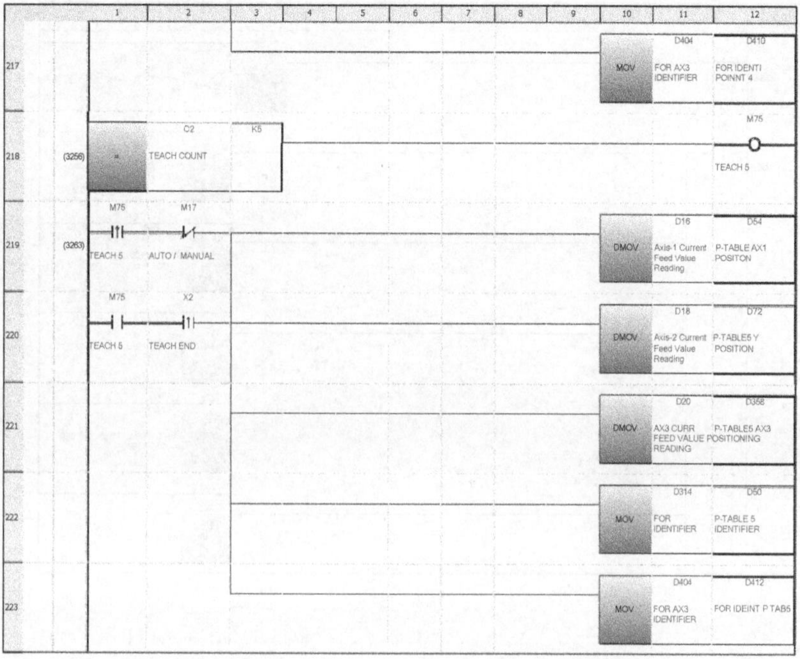
1
2
3
4
5
6
7
8
9
10
11
12
217
MOV
D404
FOR AX3 IDENTIFIER
D410
FOR IDENTI POINNT 4
218
(3256)
=
C2
TEACH COUNT
K5
M75
TEACH 5
219
(3263)
M75
TEACH 5
M17
AUTO / MANUAL
DMOV
D16
Axis-1 Current Feed Value Reading
D54
P-TABLE AX1 POSITON
220
M75
TEACH 5
X2
TEACH END
DMOV
D18
Axis-2 Current Feed Value Reading
D72
P-TABLE5 Y POSITION
221
DMOV
D20
AX3 CURR FEED VALUE READING
D358
P-TABLE5 AX3 POSITIONING
222
MOV
D314
FOR IDENTIFIER
D50
P-TABLE 5 IDENTIFIER
223
MOV
D404
FOR AX3 IDENTIFIER
D412
FOR IDEINT P TAB5

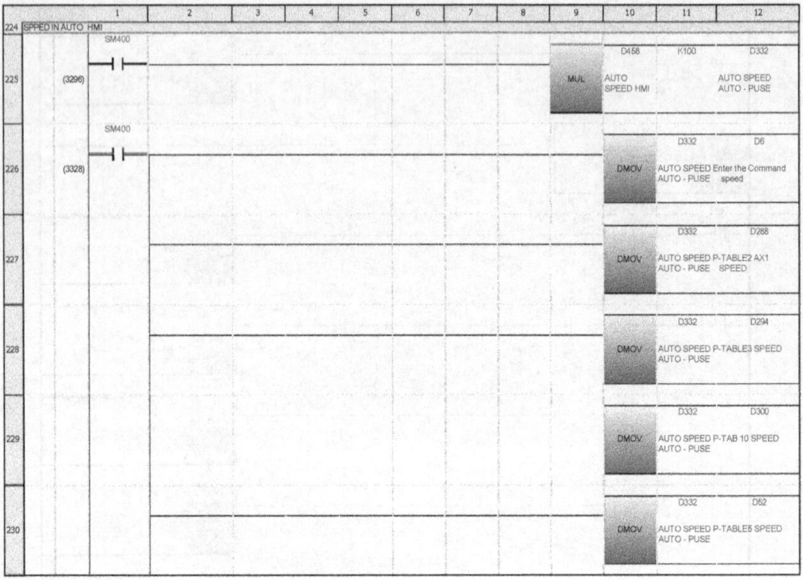
1
2
3
4
5
6
7
8
9
10
11
12
224 SPPED IN AUTO HMI
225 (3296) SM400
MUL
D458
AUTO
SPEED HMI
K100
D332
AUTO SPEED
AUTO - PUSE
226 (3328) SM400
DMOV
D332
AUTO SPEED
AUTO - PUSE
D6
Enter the Command
speed
227
DMOV
D332
AUTO SPEED
AUTO - PUSE
D288
P-TABLE2 AX1
SPEED
228
DMOV
D332
AUTO SPEED
AUTO - PUSE
D294
P-TABLE3 SPEED
229
DMOV
D332
AUTO SPEED
AUTO - PUSE
D300
P-TAB 10 SPEED
230
DMOV
D332
AUTO SPEED
AUTO - PUSE
D52
P-TABLE5 SPEED

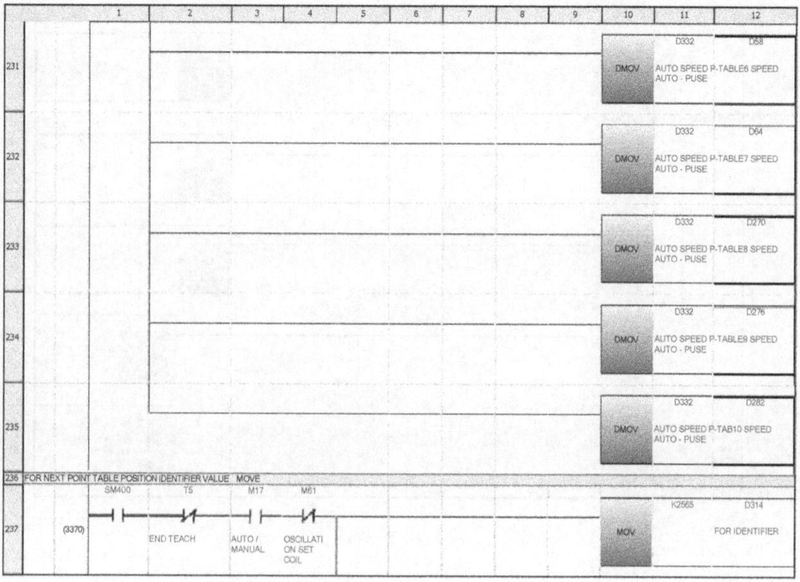
231
DMOV
D332
D58
AUTO SPEED AUTO - PUSE
P-TABLE6 SPEED
232
DMOV
D332
D64
AUTO SPEED AUTO - PUSE
P-TABLE7 SPEED
233
DMOV
D332
D270
AUTO SPEED AUTO - PUSE
P-TABLE8 SPEED
234
DMOV
D332
D276
AUTO SPEED AUTO - PUSE
P-TABLE9 SPEED
235
DMOV
D332
D282
AUTO SPEED AUTO - PUSE
P-TAB10 SPEED
236 FOR NEXT POINT TABLE POSITION IDENTIFIER VALUE MOVE
237
(3370)
SM400
T5
END TEACH
M17
AUTO / MANUAL
M61
OSCILLATION SET COIL
MOV
K2565
D314
FOR IDENTIFIER

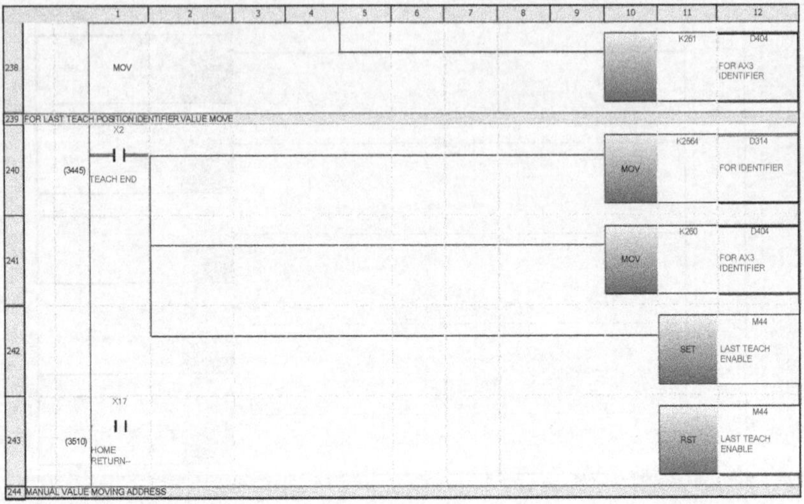
1 2 3 4 5 6 7 8 9 10 11 12
238
MOV
K261
D404
FOR AX3 IDENTIFIER
239 FOR LAST TEACH POSITION IDENTIFIER VALUE MOVE
240
(3445)
X2
TEACH END
MOV
K2564
D314
FOR IDENTIFIER
241
MOV
K260
D404
FOR AX3 IDENTIFIER
242
SET
M44
LAST TEACH ENABLE
243
(3510)
X17
HOME RETURN--
RST
M44
LAST TEACH ENABLE
244 MANUAL VALUE MOVING ADDRESS

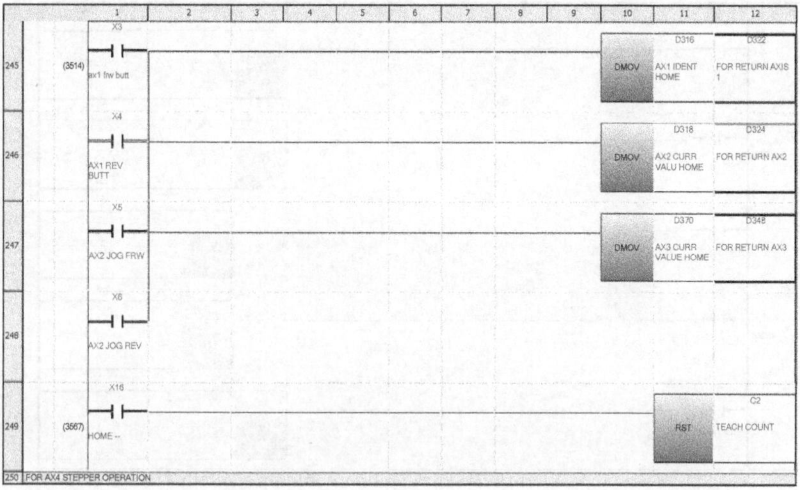
1
2
3
4
5
6
7
8
9
10
11
12
245
(3514)
X3
ax1 frw butt
DMOV
D316
AX1 IDENT HOME
D322
FOR RETURN AXIS 1
246
X4
AX1 REV BUTT
DMOV
D318
AX2 CURR VALU HOME
D324
FOR RETURN AX2
247
X5
AX2 JOG FRW
DMOV
D370
AX3 CURR VALUE HOME
D348
FOR RETURN AX3
248
X6
AX2 JOG REV
249
(3567)
X16
HOME --
RST
C2
TEACH COUNT
250
FOR AX4 STEPPER OPERATION

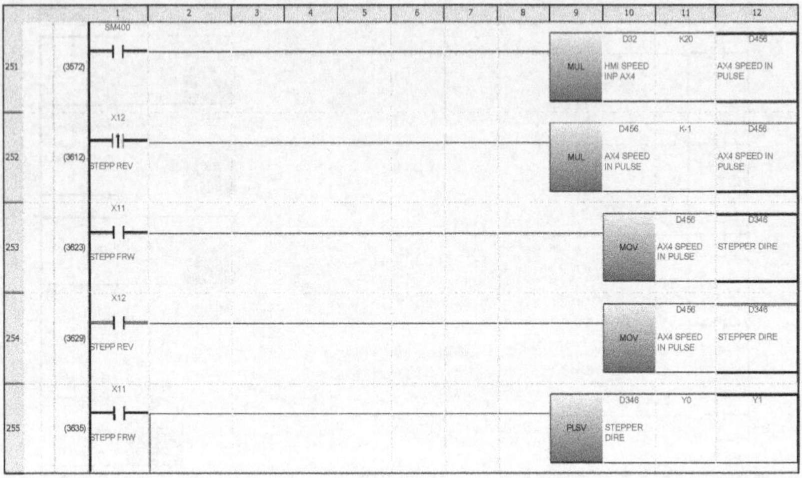
1
2
3
4
5
6
7
8
9
10
11
12
251
(3572)
SM400
MUL
D32
HMI SPEED INP AX4
K20
D456
AX4 SPEED IN PULSE
252
(3612)
X12
STEPP REV
MUL
D456
AX4 SPEED IN PULSE
K-1
D456
AX4 SPEED IN PULSE
253
(3623)
X11
STEPP FRW
MOV
D456
AX4 SPEED IN PULSE
D346
STEPPER DIRE
254
(3629)
X12
STEPP REV
MOV
D456
AX4 SPEED IN PULSE
D346
STEPPER DIRE
255
(3635)
X11
STEPP FRW
PLSV
D346
STEPPER DIRE
Y0
Y1

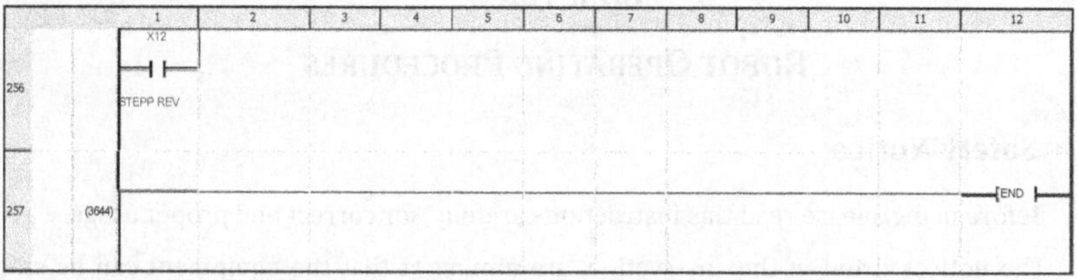
1
2
3
4
5
6
7
8
9
10
11
12
X12
256
STEPP REV
257
(3644)
END

CHAPTER 7

ROBOT OPERATING PROCEDURES

7.1. Safety Notice

- Before using, please read this instruction carefully, for correct and proper usage
- The notices listed in this instruction, are aiming at that the equipment can be safely used, and that you and other people won't get harmed or hurt.
- The design and production of this equipment has given full consideration about its safety, therefore, during the usage, please do obey the notices listed in this instruction, otherwise, there may cause harmful incidents to its users or other concerning people.

7.1.1. Safety Notices that Need to be Observed

- Construction of the input power supply, selection of the configuration site, usage of high pressure gases, safekeeping and configuration, safekeeping of welded work pieces, disposal of wastes, etc., please observe concerning regulations and the internal standards of your esteemed company
- Irrelevant personnel are forbidden to enter the equipment operation site. The magnetic field created when the power supply of a welding machine is connected may cause adverse effects to the functions of heart starters.
- Professionals and qualified personnel are required to carry out the installation, repairs and maintenance of the equipment.
- For safety concern, please correctly understand the contents of this instruction, and operations of the equipment are required to be carried out by personnel who have safety usage knowledge and concerning skills.
- Please do not touch the electrified parts
- Concerning electricians are required to wire the equipment in strict compliance with requirements of the equipment on power supply, voltage or frequency and the power supply for welding work, and the base material shall be grounded reliably.
- Please do not use cables whose capacity is inadequate, or whose insulation protection sheath has been damaged which leads to the result that the conductor inside has been exposed.
- It is forbidden to raise or lower the current value of each fusible core or circuit breakers.

- During hanging operation of the equipment, the hanging cable shall be tied to the suspension ring, and in order to prevent tilting and falling of the equipment, it is suggested that your esteemed company prepare concerning hanging tools that have sufficient strength.
- Regular maintenance and repairs shall be carried out, and the equipment shall not be used when there is damaged part that has not been repaired.
- When the equipment is not used, please turn off all input power supplies.
- It is suggested that keys to the electric cabinet shall be safely kept, and non-professionals are forbidden to open the electric cabinet.

Flowchart

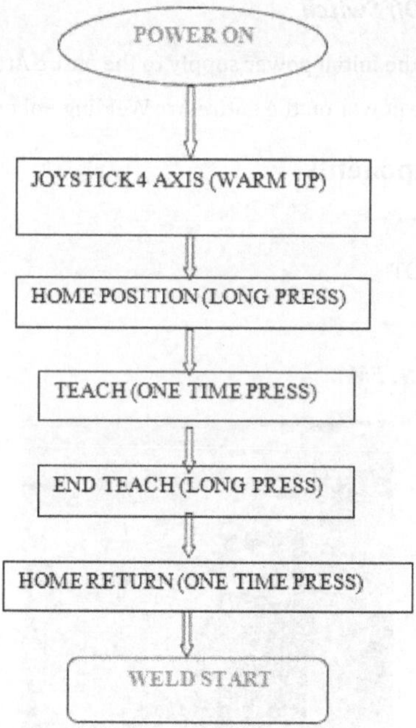

7.2. Four Axis Arc Welding Robot

- The Four axis Arc welding Robot is designed to weld the component by using remote pendant.
- The name itself defined that this Arc Welding Robot can be moved about the four axis

7.2.1. *Design Specification of Robot*

Number of axis	:	Four
Load carrying capacity	:	50 kg
Maximum rapid	:	9000mm/min
Reach and Stroke	:	2000 mm & 120 mm
Tool orientation	:	Perpendicular(Downhand)
Repeatability	:	± 0.2 mm
Precision and accuracy	:	± 0.2 mm
Operating environment	:	Hazardous (GMAW & FCAW)

7.2.2. *Main Power On/Off Switch*

- ON switch provides the initial power supply to the entire Arc Welding Robot System.
- OFF switch shuts the power on the entire Arc Welding Robot System.

7.3. Important Components

- PLC CONTROL SYSTEM
- ARC WELDING ROBOT
- REMOTE PENDANT

7.3.1. *PLC CONTROL SYSTEM*

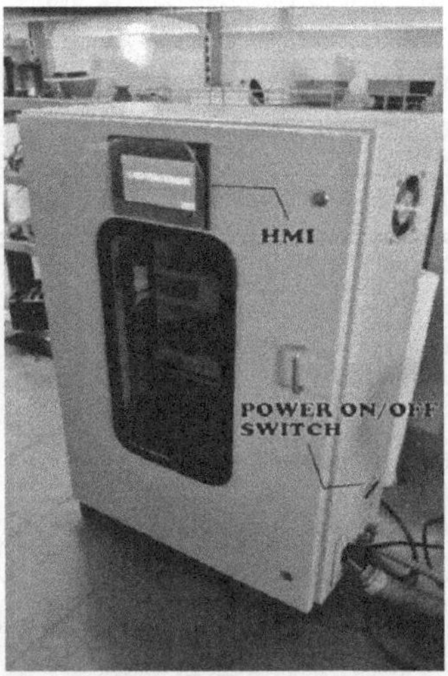

The PLC control system consists of the following components

- Driving Unit
- Programming Unit
- HMI Display Unit
- Switching Components
- Power supply unit

7.3.2. Driving Unit

It consists of Servo Amplifier, Servo Motors and Stepper Drive which are having the specifications as shown in the table.

Axis	Model/Make		Capacity
Axis-1	Servo Amplifier	MR-JE-70B	2.4 Nm, 750 W, 3000 rpm, 4.8 A
	Servo Motor	HG KR73 B	
Axis-2	Servo Amplifier	MR-JE-40B	1.3 Nm, 400 W, 3000 rpm, 2.6 A
	Servo Motor	HG KR73 B	
Axis-3	Servo Amplifier	MR-JE-40B	1.3 Nm, 400 W, 3000 rpm, 2.6 A
	Servo Motor	HG KR73 B	
Axis-4	Steppe Drive& Motor	DMA860H	3.1 Nm Holding Torque, 4.5 amps

7.3.3. Programming Unit

- This unit comprises of the following parts in it to control the entire system.
- PLC unit Model: FX5U-64MTES/UL 32di/32do
- Position Module with SSNET communication system

7.3.4. HMI Display Unit

- HMI (Human Machine Interface)
- Model : HMI GS2107WBDA
- Screen Size : 7" Graphical Touch Screen (Colour)

Home Screen

- It is a Graphical Touch Screen which shows the Machine Name & manufacturer's Name and Address
- Main - To go to the Main screen

Parameter Screen

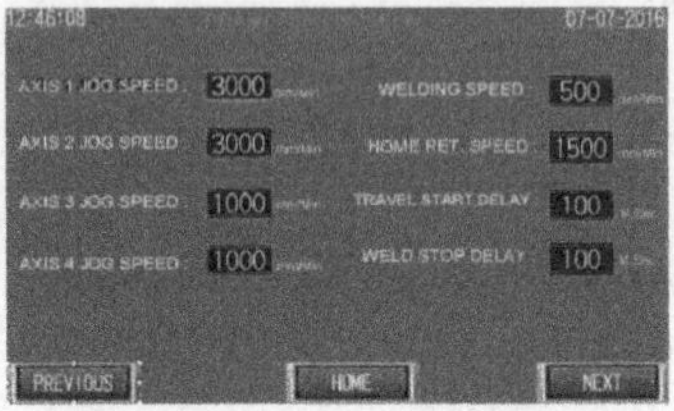

- This screen displays the information and settings of the welding parameters
- By using this touch screen display, we can enter the numerical values to set the welding speed as well as Axis speeds.
- We can see how many "Teaches (Reference points)" that we have given to define the weld path.
- Maximum 10 reference points only will be accepted by the control system.

WELDING SPEED : Travel speed of the welding Robot

HOME RET SPEED : The speed at which the Welding torch return to home position

TRAVEL START DELAY : The delay between the arc initiating and the travel starting

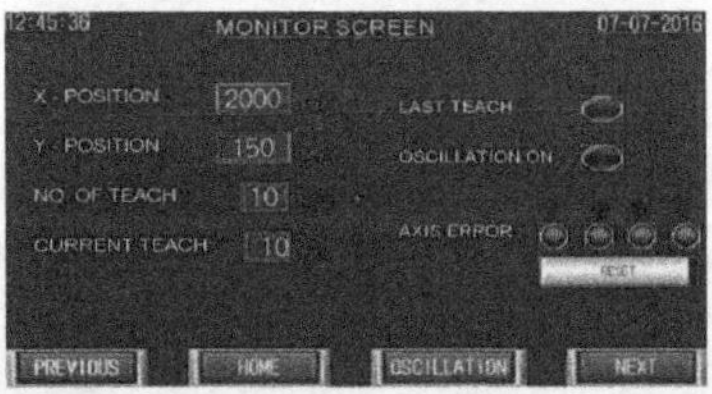

WELD STOP DELAY : The delay between travel stop and arc end

JOG SPEED : The speed of the corresponding axis movement.

HOME : To go to the Home screen

PREVIOUS : To go to the previous screen

NEXT : To go to the Next screen

Monitor Screen: This screen displays the current position of the X axis and Y axis and It also shows the Total Number of Teaches and current Teach.

Note

X-Position	: Current position of the X-axis
Y-Position	: Current position of the Y-axis
NO.OF TEACH	: Total number of Teaches in that program
LAST TEACH	: It glows when the last is given
AXIS ERROR	: It indicates the Axis error.
CURRENT TEACH	: At which Teach point the robot is currently performing the welding
OSCILLATION ON	: It shows about the oscillation movement in welding (ON/OFF)

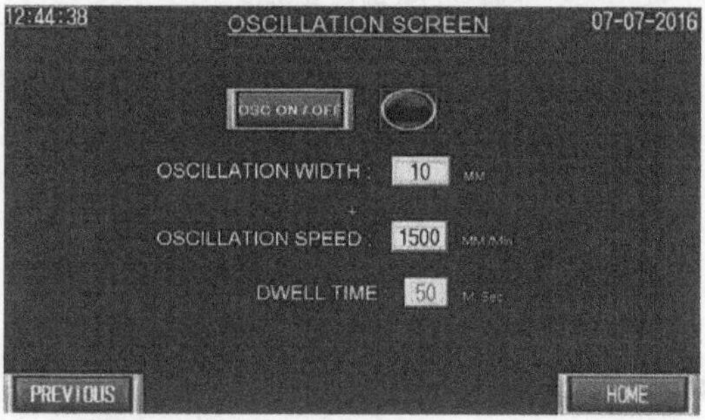

Oscillation Screen (optional)

- The screen which is named as an oscillation screen has additional feature in the machine. We can perform the welding the weaving movement.

Note

OSCILLATION WIDTH : Maximum width of the oscillation of the welding torch for widening the weld bead

OSCILLATION SPEED : It indicates the speed of the welding torch oscillation

DWELL TIME : The time to stay on the edges of weld bead in micro seconds.

Switching Components

- The switching system is assembled by using the MCCB, MCB and Fuses.

Power Supply Unit

It consists of an Isolation Transformer (5KVA) and Regulated power supply (24 V/ 5 A).

7.3.5. *Arc Welding Robot*

- Strictly the Arc Welding Robot has to be fixed at extremely close to the weld joint.

Four Axis

- Longitudinal Axis
- Cross slide Axis
- Vertical Axis
- Rotational Axis (PITCH)

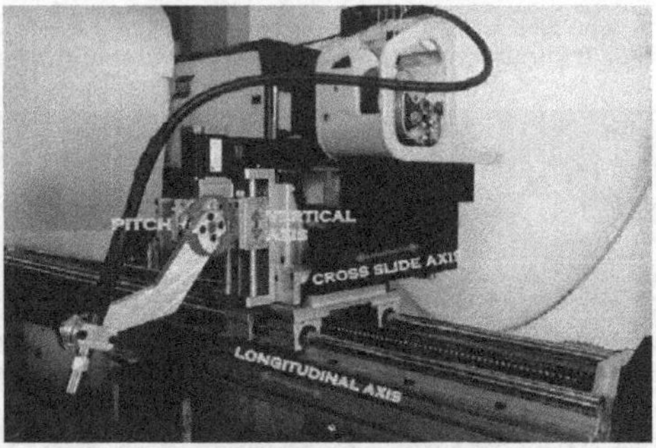

7.3.6. *Specifications*

Axis	Stroke Length (Max) mm	Travel Speed mm/min	
		Min	Max
Longitudinal Axis	2000	30	9000
Cross Slide Axis	150	30	4500
Vertical Axis	70	30	3000
Rotational Axis	180°	30	1200

7.3.7. *Remote Pendant*

Remote pendant is a device which is used to program the welding path by assigning the reference points in Teach Mode.

7.3.8. Joystick-1

Control of Longitudinal Axis and Cross Slide Movement

Longitudinal Movement Control

By Sliding the Joystick-1 to your right or left hand side, the arc welding robot can be moved in the longitudinal direction.

Cross Slide Movement Control

The Arc Welding Robot can be moved in to & fro Cross Slide Axis by means of moving the Joystick-1 in the up & down direction.

7.3.9. Joystick–2

Control of Vertical Axis and Rotational Axis (Pitch)

Vertical Movement Control

- By Moving the Joystick-2 up and down, the distance between torch tip and the work piece (vertical movement) can be adjusted.

Rotational Axis Control

- By Sliding the Joystick-2 to your right or left hand side, the welding torch angle i.e. Rotational axis (Pitch) can be adjusted.

Joystick	Direction	Movement
Joystick-1	→	Robot Movement **to right hand side**
	←	Robot Movement **to left hand side**
	↑	**Cross Slide Backward** Movement
	↓	**Cross slide Forward** Movement
Joystick-2	→	**Welding Torch** Angle movement in **Downward direction**
	←	**Welding Torch** Angle movement in **upward direction**
	↑	**Vertical** movement in **upward direction**
	↓	**Vertical** movement in **downward direction**

7.3.10. Home

- HOME position is the **Starting Point** of the welding.
- After positioned the welding torch where you want to start the weld, that point can be set as "HOME" position by pressing the button till the **light glow.**
- The same button acts a **Return to Home button.**

7.3.11. Teach

- TEACH button is an important key to fix or **to teach the weld path to the Robot.**
- After Selecting the HOME position, immediately TEACH button will glow to indicate that it is ready for getting the input values.
- At every teach point, we have to press the button till the Green light off which is an acceptance of TEACH point.
- For the **FINAL TEACH POINT**, we have to give a **LONG PRESS.** At the same time HOME button will glow.
- Then Press the HOME button to bring back the Welding Torch to its HOME Position.
- We can provide maximum 10 teach points to the robot for single welding program.

7.3.12. Start

- START is used to indicate the Start of welding.
- After Positioning the Welding Torch in its HOME position, the Start Button Will Glow. This is an indication to start the welding.

7.3.13. Stop

- This STOP button acts as **an emergency weld cycle stop.**
- If STOP button glows, it indicates that the Arc Welding **Robot reaches its maximum movement in that particular direction.**
- Once the **STOP glows**, the joystick movement will be seized unless reset the STOP button AGAIN.

7.4. Example Program

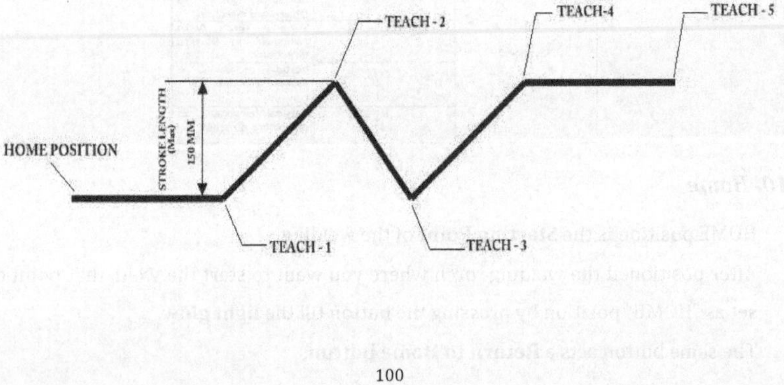

7.4.1. *Procedure*

Step 1

- Switch ON the PLC control system and also Welding Power Source.

Step 2

- You have to move JOYSTICK-1 & JOYSTICK-2 in the appropriate direction to reach the exact position where the welding is to be initiated.

Step 3

- After reached the exact position, Press the HOME button till it glows. This point is the HOME position for the welding operation.

Step 4

- From the HOME position, you can move the welding robot arbitrarily to reach the "TEACH-1" by using JOYSTICK-1 and JOYSTICK-2.
- The TEACH-1 is the first reference point for the robot to get along with the weld path.
- To select "TEACH-1" as a reference point, Press the TEACH button till the light off for fewer seconds.

Step 5

- From TEACH-1 point, you can move the welding robot arbitrarily to reach the "TEACH-2" by using JOYSTICK-1 and JOYSTICK-2.
- To select "TEACH-2" as a second reference point, Press the TEACH button till the light off for fewer seconds.

Step 6

- Repeat the STEP 5 for the remaining TEACH points, say TEACH-3, TEACH-4 and TEACH-5.
- For the "final TEACH" i.e TEACH-5 , long press is required on TEACH button till the home glow.

Step 7

- Press the HOME button to bring back the welding robot to its HOME position.

Step 8

- Once the welding torch arrive the HOME position , the start button will glow. This is an indication that the robot is READY to perform the welding in the pre-defined path.

- Press the start button to perform the welding.

Step 9

- At end of the welding, welding robot will be remained at the last Teach point, i.e. TEACH-5.
- Press the HOME button to bring the torch welding robot to its HOME position.

Step10

- Repeat the Step 9 for the deposition of additional layer to complete the welding.

CHAPTER 8

ROBOT INSTALLATION PROCEDURES

Four Axis Arc Welding Robot

The Four axis Arc welding Robot is designed to weld the component by using remote pendant.

The name itself defined that this Arc Welding Robot can be moved about the four axis

8.1. Design Specification of Robot

Number of axis	:	Four
Load carrying capacity	:	50 kg
Maximum rapid	:	9000mm/min
Reach and Stroke	:	2000 mm & 120 mm
Tool orientation	:	perpendicular
Repeatability	:	± 0.2 mm
Precision and accuracy	:	± 0.2 mm
Operating environment	:	Hazardous (GMAW & FCAW)

8.2. Main Power On/Off Switch

ON switch provides the initial power supply to the entire Arc Welding Robot System.

OFF switch shuts the power on the entire Arc Welding Robot System.

8.3. Important Components

PLC CONTROL SYSTEM

ARC WELDING ROBOT

REMOTE PENDANT

8.4. PLC Control System

The PLC control system consists of the following components

- Driving Unit
- Programming Unit
- HMI Display Unit
- Switching Components
- Power supply unit

8.4.1. Driving Unit

It consists of Servo Amplifier, Servo Motors and Stepper Drive which are having the specifications as shown in the table.

Axis	Model/Make		Capacity
Axis-1	Servo Amplifier	MR-JE-70B	2.4 Nm, 750 W, 3000 rpm, 4.8 A
	Servo Motor	HG KR73 B	
Axis-2	Servo Amplifier	MR-JE-40B	1.3 Nm, 400 W, 3000 rpm, 2.6 A
	Servo Motor	HG KR73 B	
Axis-3	Servo Amplifier	MR-JE-40B	1.3 Nm, 400 W, 3000 rpm, 2.6 A
	Servo Motor	HG KR73 B	
Axis-4	Steppe Drive& Motor	DMA860H	3.1 Nm Holding Torque, 4.5 amps

8.4.2. Programming Unit

- This unit comprises of the following parts in it to control the entire system.
- PLC unit Model: FX5U-64MTES/UL 32di/32do
- Position Module with SSNET communication system

8.5. Installation Procedure

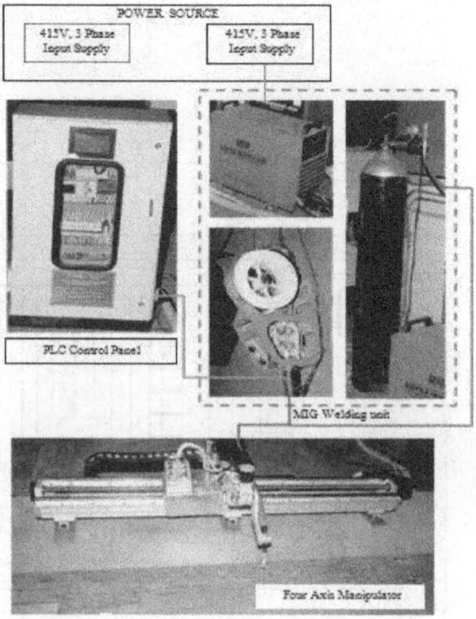

Installation of Four Axis Arc Welding Robot- Photographic Views

8.6. Block Diagram

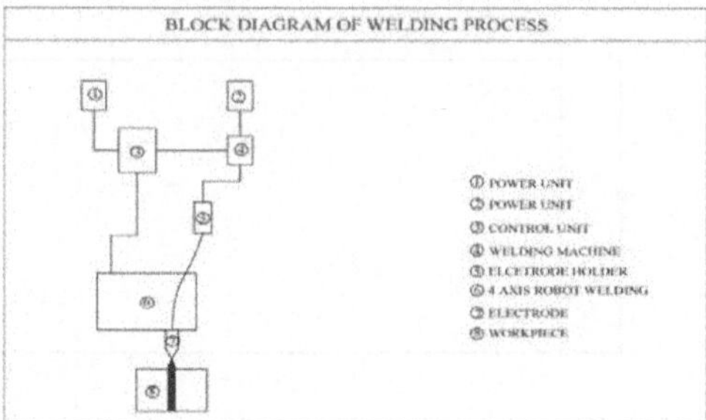

The Wiring diagram of control system are attached as Appendix 'A' as follows.

Sl No	Description	Remarks
01	Control panel layout diagram	
02	Terminal wiring connection diagram	
03	PLC output wiring diagram	
04	24V Circuit wiring diagram	
05	Power circuit wiring diagram - 1	
06	Power circuit wiring diagram - 2	
07	System Diagram	

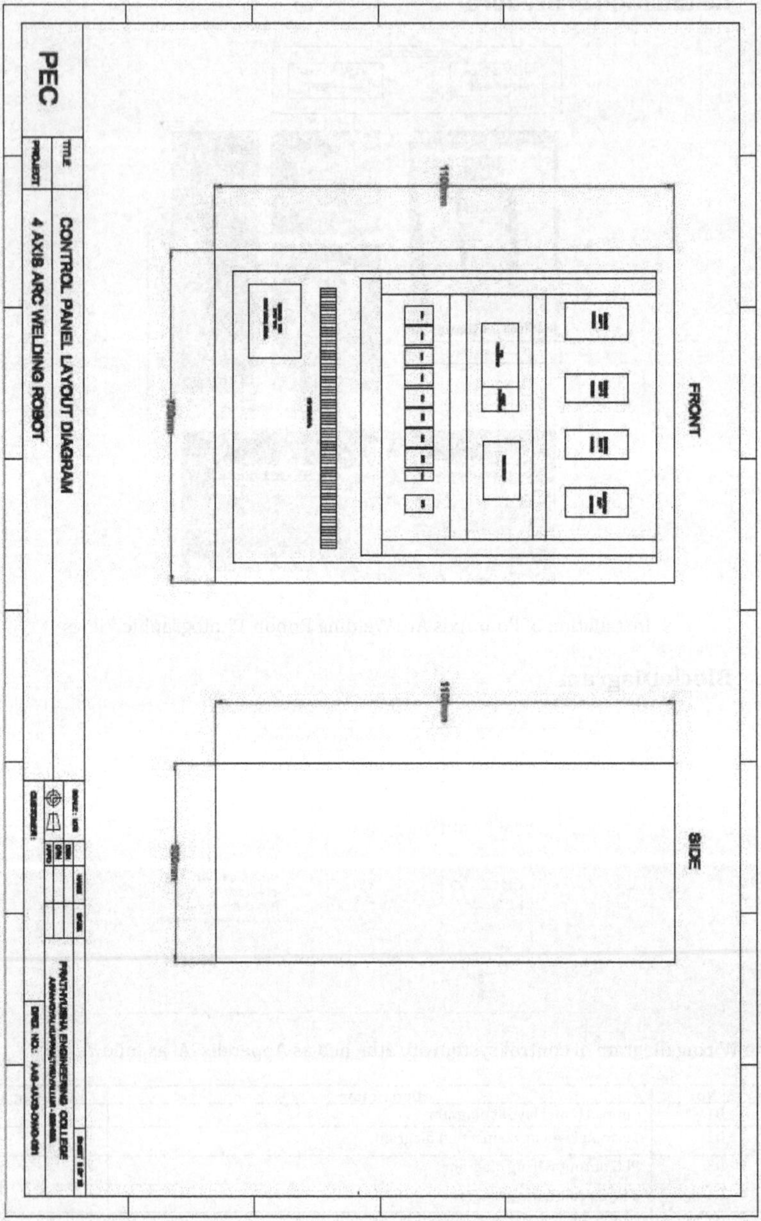

PEC
TITLE
CONTROL PANEL LAYOUT DIAGRAM
PROJECT
4 AXIS ARC WELDING ROBOT
FRONT
SIDE
1100mm
1100mm
700mm
PRATHYUSHA ENGINEERING COLLEGE

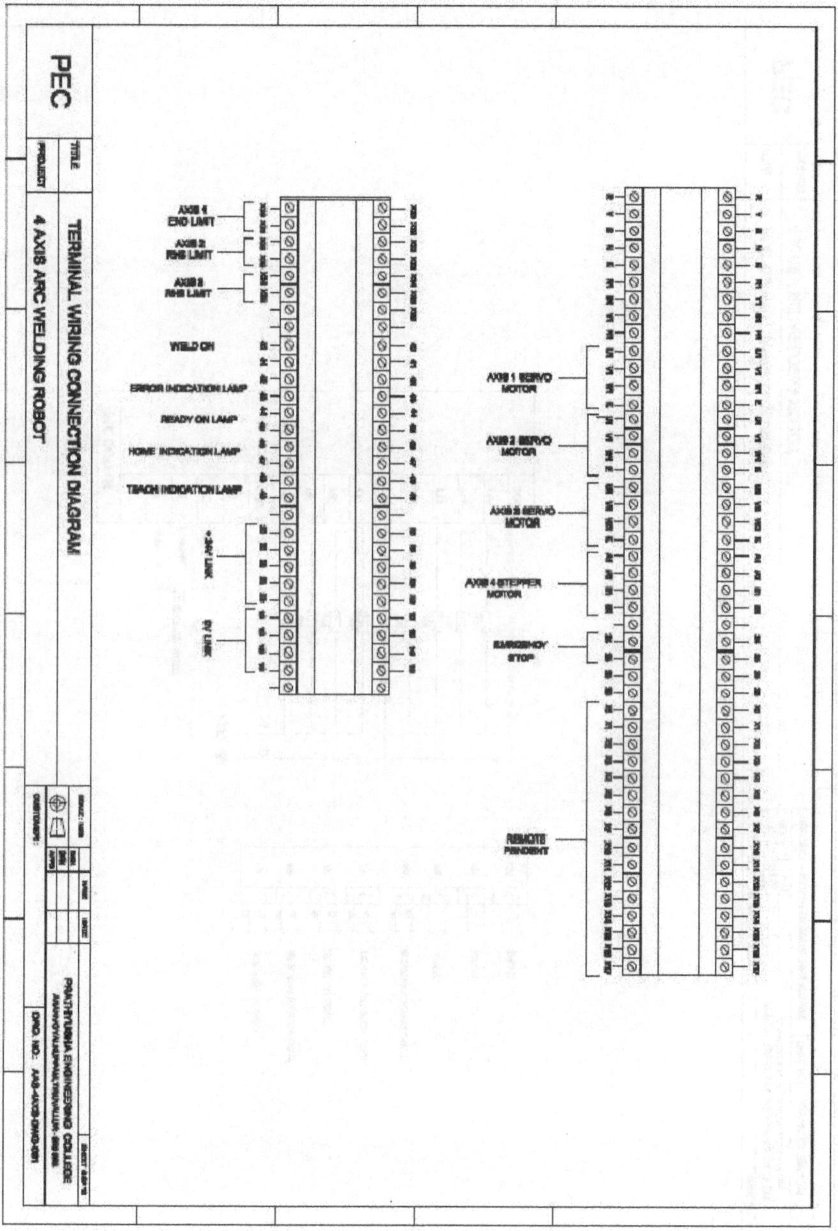
PEC
TITLE
TERMINAL WIRING CONNECTION DIAGRAM
PROJECT
4 AXIS ARC WELDING ROBOT
AXIS 1 END LIMIT
AXIS 2 RHS LIMIT
AXIS 3 RHS LIMIT
WELD ON
ERROR INDICATION LAMP
READY ON LAMP
HOME INDICATION LAMP
TEACH INDICATION LAMP
+24V LINK
0V LINK
AXIS 1 SERVO MOTOR
AXIS 2 SERVO MOTOR
AXIS 3 SERVO MOTOR
AXIS 4 STEPPER MOTOR
EMERGENCY STOP
REMOTE PENDENT
PRATHYUSHA ENGINEERING COLLEGE

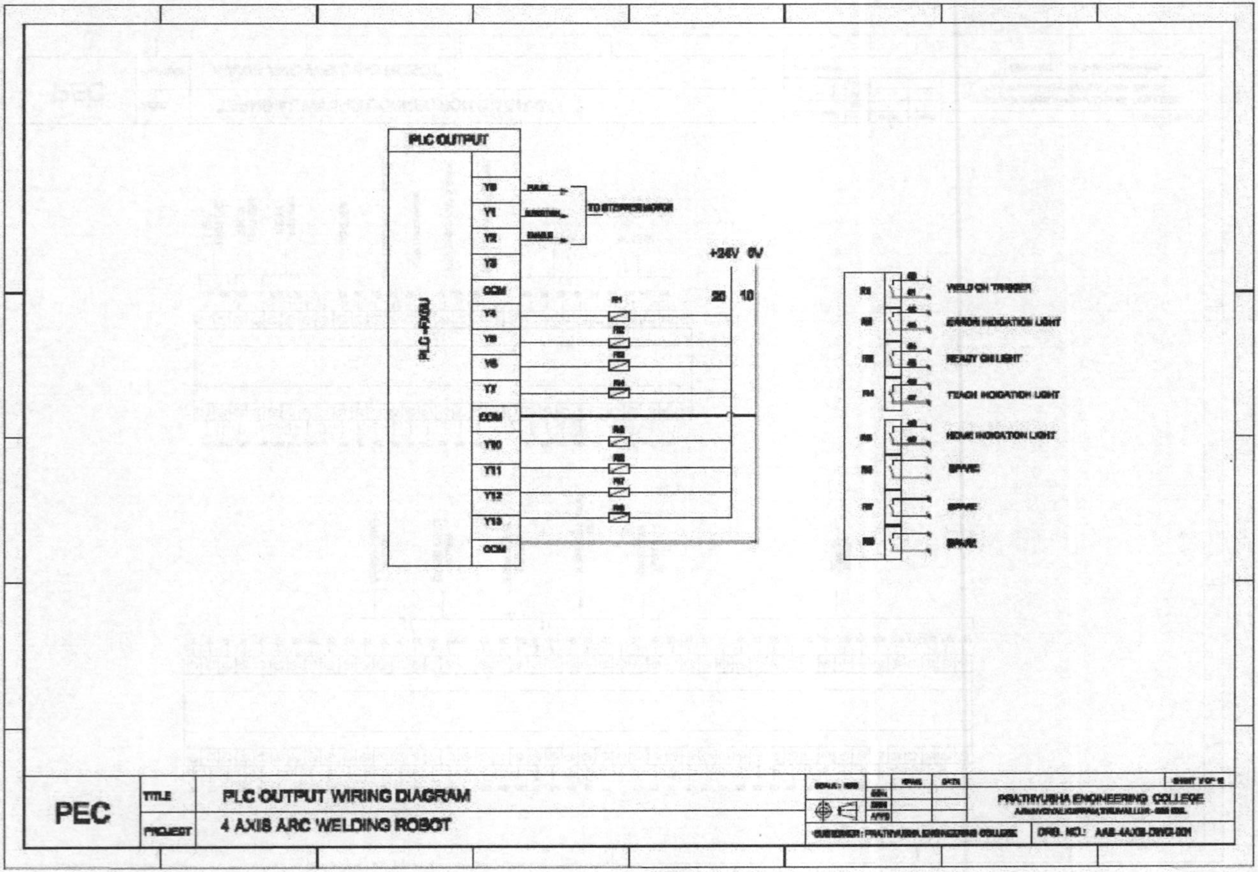

PLC OUTPUT
Y0
Y1
Y2
Y3
COM
Y4
Y5
Y6
Y7
COM
Y10
Y11
Y12
Y13
COM
TO STEPPER MOTOR
+24V 0V
20 10
WELD ON TRIGGER
ERROR INDICATION LIGHT
READY ON LIGHT
TEACH INDICATION LIGHT
HOME INDICATION LIGHT
SPARE
SPARE
SPARE
PEC
TITLE
PLC OUTPUT WIRING DIAGRAM
PROJECT
4 AXIS ARC WELDING ROBOT
PRATHYUSHA ENGINEERING COLLEGE

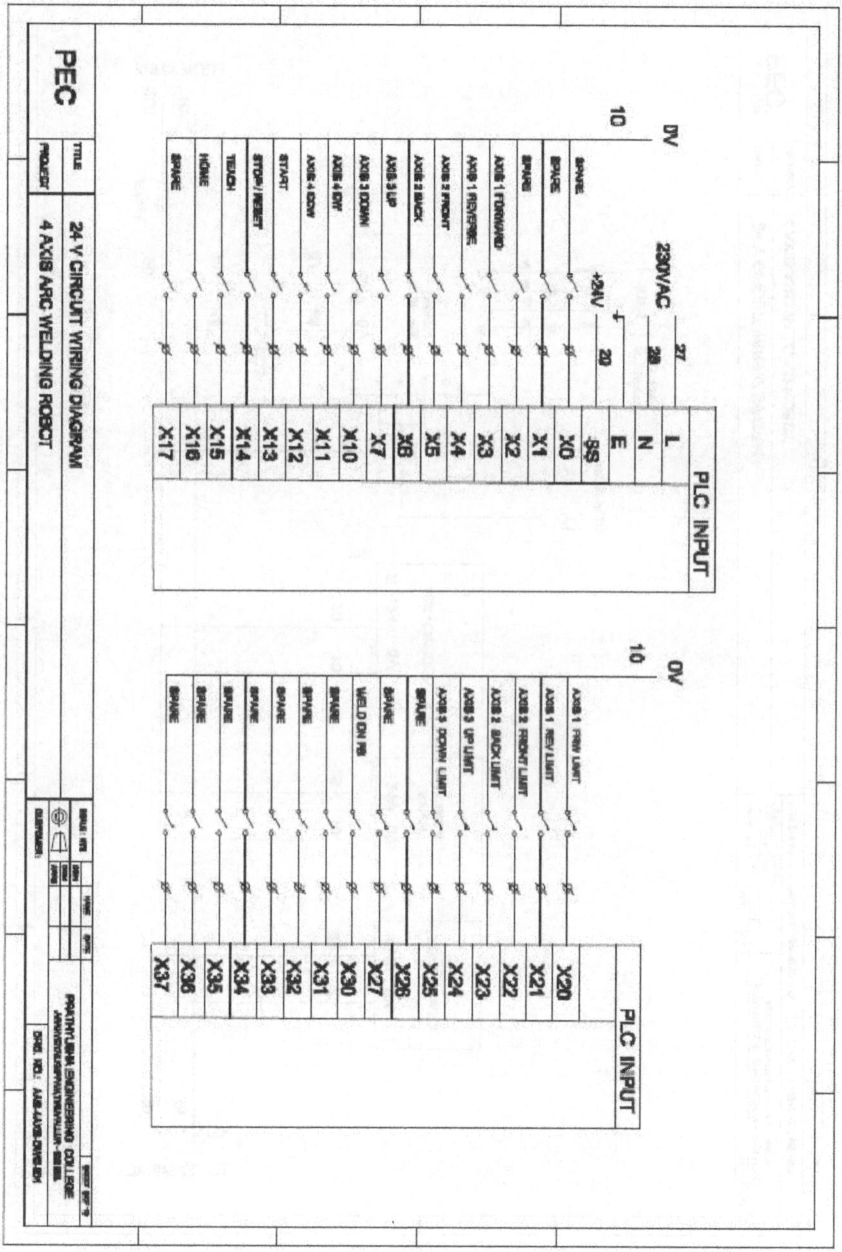
PLC INPUT
0V
10
230VAC
+24V
27
28
20
L
N
E
S/S
SPARE
SPARE
SPARE
AXIS 1 FORWARD
AXIS 1 REVERSE
AXIS 2 FRONT
AXIS 2 BACK
AXIS 3 UP
AXIS 3 DOWN
AXIS 4 CW
AXIS 4 CCW
START
STOP / RESET
TEACH
HOME
SPARE
X0
X1
X2
X3
X4
X5
X6
X7
X10
X11
X12
X13
X14
X15
X16
X17
PLC INPUT
0V
10
AXIS 1 FRW LIMIT
AXIS 1 REV LIMIT
AXIS 2 FRONT LIMIT
AXIS 2 BACK LIMIT
AXIS 3 UP LIMIT
AXIS 3 DOWN LIMIT
SPARE
SPARE
WELD ON PB
SPARE
SPARE
SPARE
SPARE
SPARE
SPARE
SPARE
X20
X21
X22
X23
X24
X25
X26
X27
X30
X31
X32
X33
X34
X35
X36
X37
PEC
TITLE
24 V CIRCUIT WIRING DIAGRAM
PROJECT
4 AXIS ARC WELDING ROBOT
PRATHYUSHA ENGINEERING COLLEGE

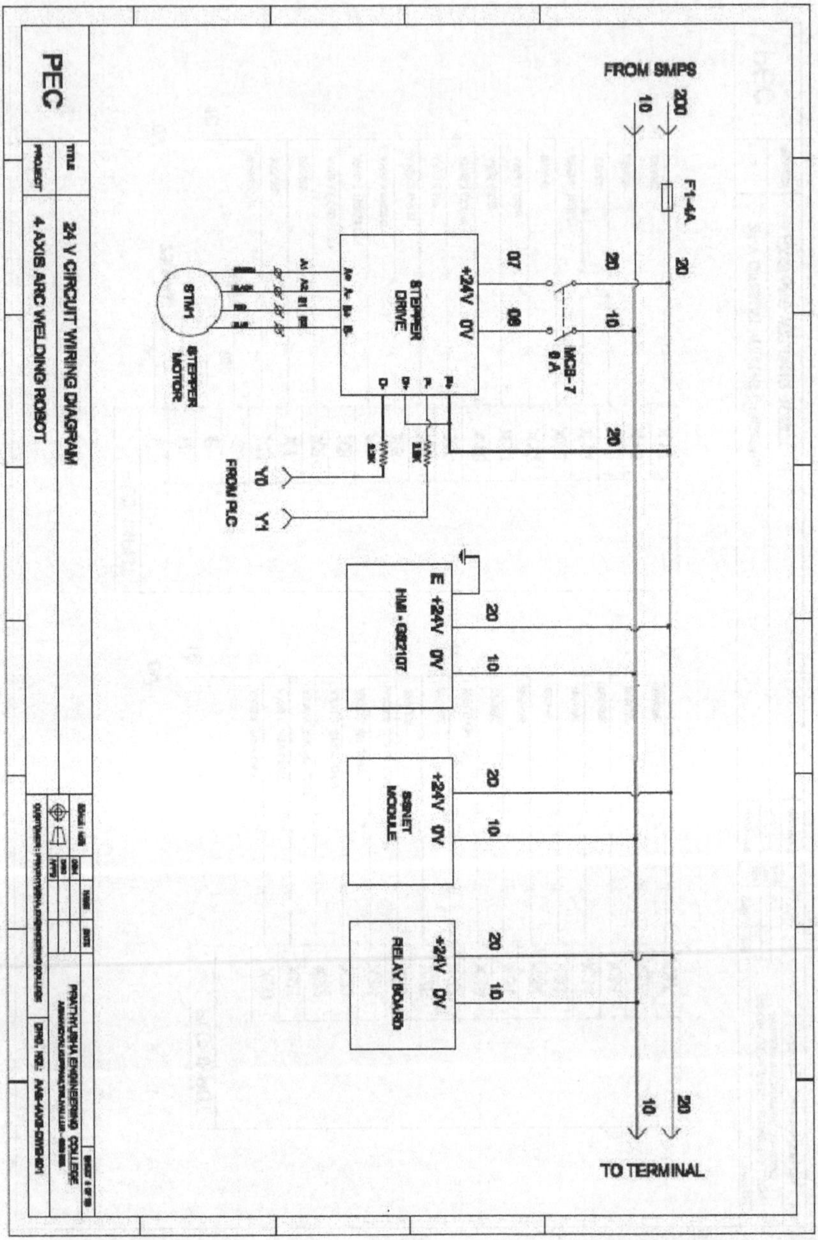
FROM SMPS
200
10
F1-4A
20
20
10
MCB-7
6 A
07
08
+24V
0V
STEPPER DRIVE
STM1
STEPPER MOTOR
FROM PLC
Y0
Y1
E
+24V
0V
HMI - OB210T
SSNET MODULE
RELAY BOARD
20
10
TO TERMINAL
PEC
TITLE
24 V CIRCUIT WIRING DIAGRAM
PROJECT
4 AXIS ARC WELDING ROBOT
PRATHYUSHA ENGINEERING COLLEGE

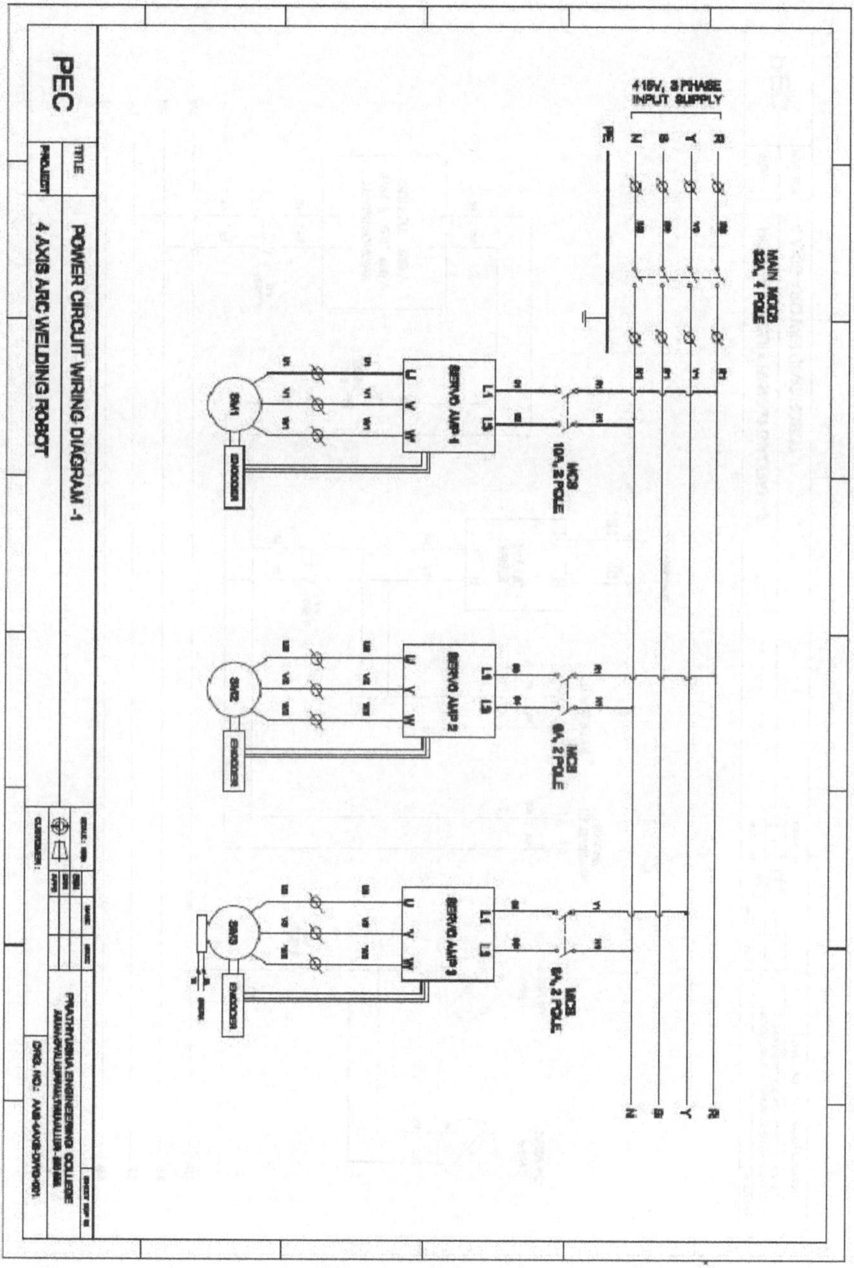
PEC
TITLE
POWER CIRCUIT WIRING DIAGRAM -1
PROJECT
4 AXIS ARC WELDING ROBOT
415V, 3 PHASE INPUT SUPPLY
MAIN MCCB 32A, 4 POLE
R
Y
B
N
PE
MCB 10A, 2 POLE
MCB 6A, 2 POLE
SERVO AMP 1
SERVO AMP 2
SERVO AMP 3
L1
L3
U
V
W
SM1
SM2
SM3
ENCODER
PRATHYUSHA ENGINEERING COLLEGE

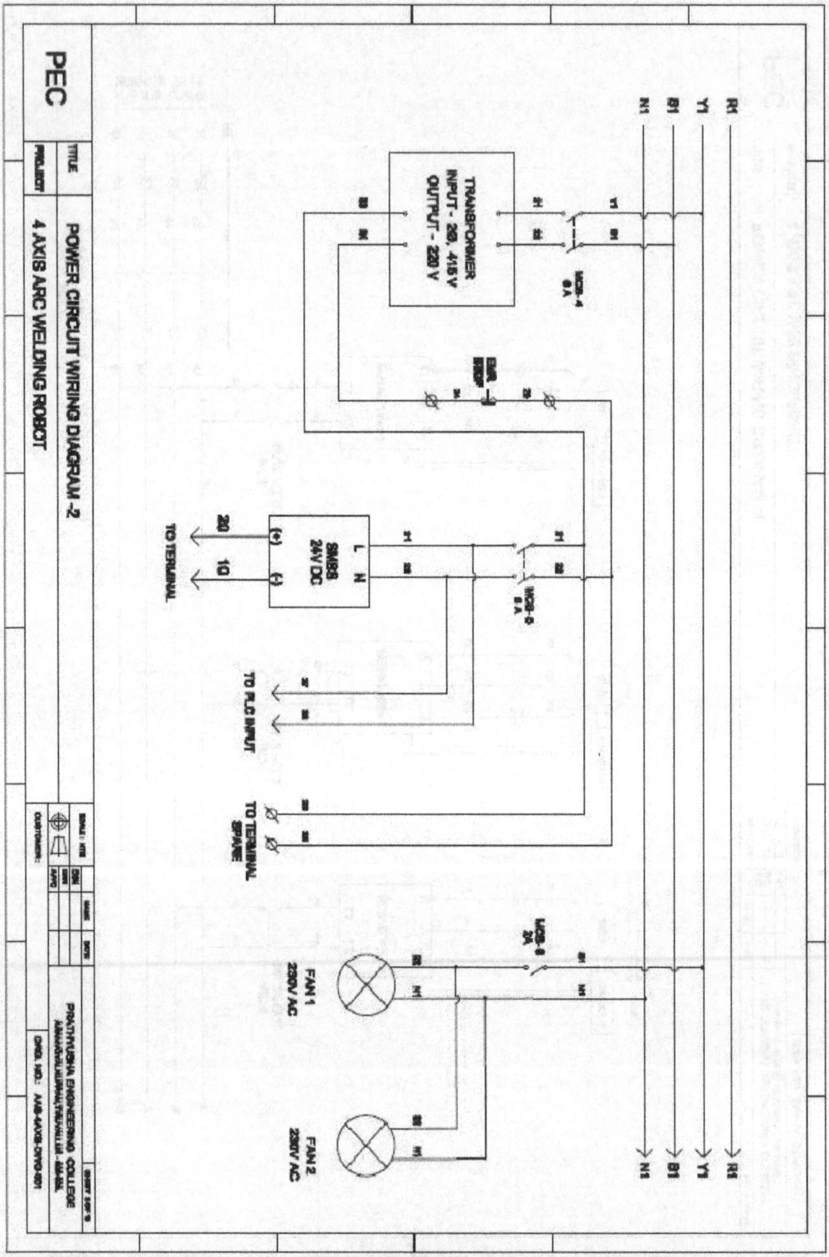
PEC
TITLE
POWER CIRCUIT WIRING DIAGRAM -2
PROJECT
4 AXIS ARC WELDING ROBOT
PRATHYUSHA ENGINEERING COLLEGE
R1
Y1
B1
N1
TRANSFORMER
SMPS
24V DC
TO TERMINAL
TO PLC INPUT
TO TERMINAL
SPARE
FAN 1
230V AC
FAN 2
230V AC

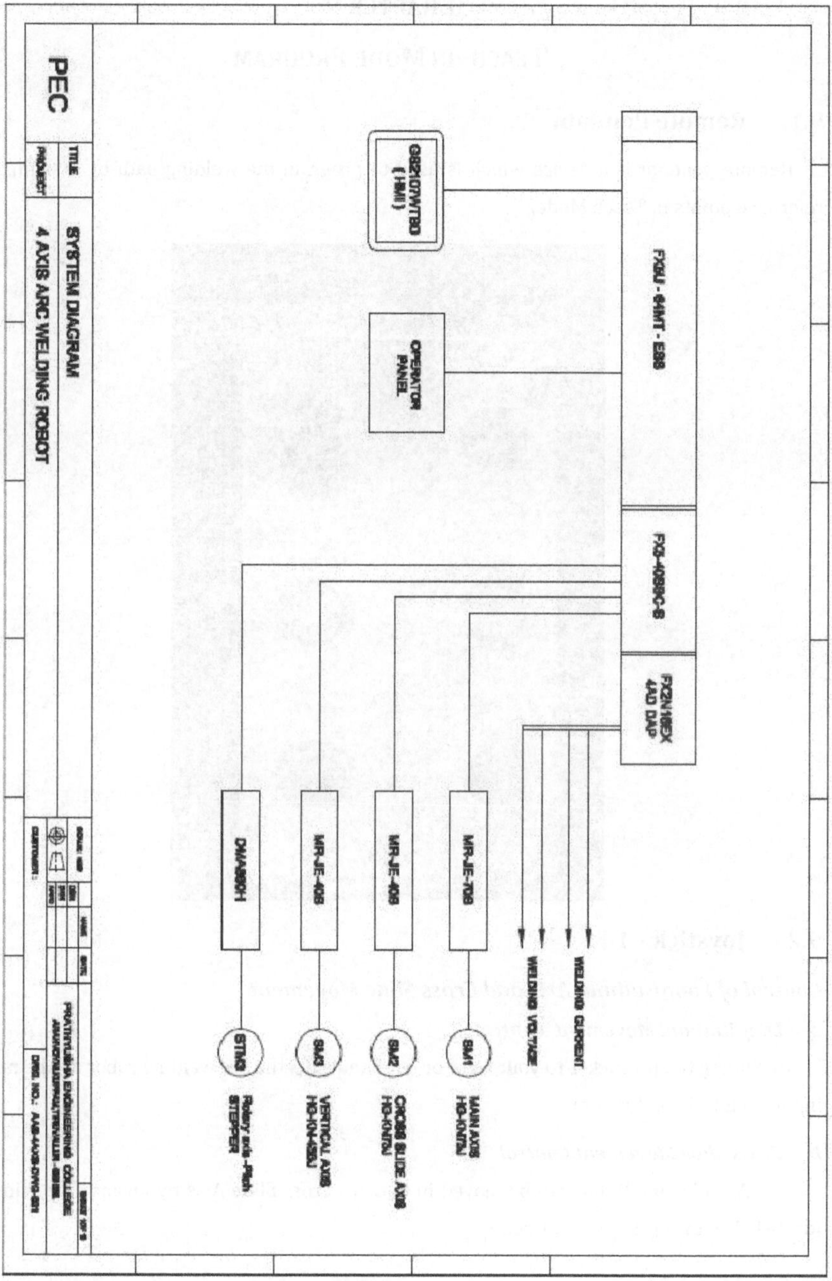
FX5U - 64MT - ESS
FX5-40SSC-S
FX2N16EX
4AD DAP
GS2107WTBD
(HMI)
OPERATOR
PANEL
WELDING CURRENT
WELDING VOLTAGE
MR-JE-70B
MR-JE-40B
MR-JE-40B
DMA860H
SM1
SM2
SM3
STM3
MAIN AXIS
HG-KN73J
CROSS SLIDE AXIS
HG-KN73J
VERTICAL AXIS
Rotary axis -Pitch
STEPPER
PEC
TITLE
SYSTEM DIAGRAM
PROJECT
4 AXIS ARC WELDING ROBOT
NAME
DATE
PRATHYUSHA ENGINEERING COLLEGE
CUSTOMER :

CHAPTER 9

TEACH–IN MODE PROGRAM

9.1. Remote Pendant

Remote pendant is a device which is used to program the welding path by assigning the reference points in Teach Mode.

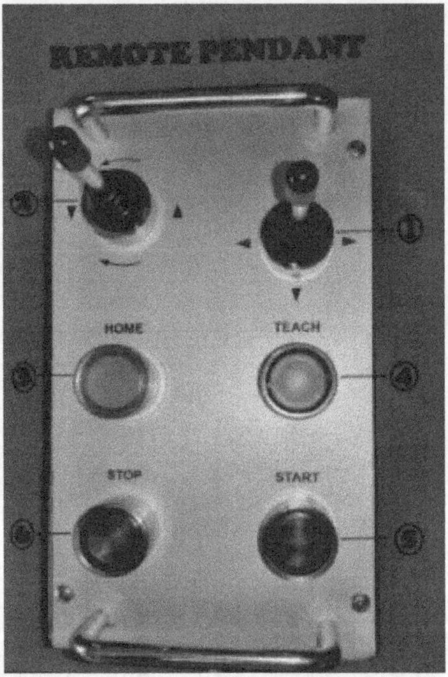

9.2. Joystick - 1

Control of Longitudinal Axis and Cross Slide Movement

a. Longitudinal Movement Control

By Sliding the Joystick-1 to your right or left hand side, the arc welding robot can be moved in the longitudinal direction.

b. Cross Slide Movement Control

The Arc Welding Robot can be moved in to & fro Cross Slide Axis by means of moving the Joystick-1 in the up & down direction.

9.3. Joystick - 2

Control of Vertical Axis and Rotational Axis (Pitch)

a. Vertical Movement Control

By Moving the Joystick-2 up and down, the distance between torch tip and the work piece (vertical movement) can be adjusted.

b. Rotational Axis Control

By Sliding the Joystick-2 to your right or left hand side, the welding torch angle i.e. Rotational axis (Pitch) can be adjusted.

Joystick	Direction	Movement
Joystick-1	→	Robot Movement to right hand side
	←	Robot Movement to left hand side
	↑	Cross Slide Backward Movement
	↓	Cross slide Forward Movement
Joystick-2	→	Welding Torch Angle movement in Downward direction
	←	Welding Torch Angle movement in upward direction
	↑	Vertical movement in upward direction
	↓	Vertical movement in downward direction

9.4. Home

- HOME position is the **Starting Point** of the welding.
- After positioned the welding torch where you want to start the weld, that point can be set as "HOME" position by pressing the button till the **light glow.**
- The same button acts a **Return to Home button**.

9.5. Teach

- TEACH button is an important key to fix or **to teach the weld path to the Robot.**
- After Selecting the HOME position, immediately TEACH button will glow to indicate that it is ready for getting the input values.
- At every teach point, we have to press the button till the Green light off which is an acceptance of TEACH point.
- For the **FINAL TEACH POINT**, we have to give a **LONG PRESS.** At the same time HOME button will glow.
- Then Press the HOME button to bring back the Welding Torch to its HOME Position.
- We can provide maximum 10 teach points to the robot for single welding program.

9.6. Start

- START is used to indicate the Start of welding.
- After Positioning the Welding Torch in its HOME position, the Start
- Button Will Glow. This is an indication to start the welding.

9.7. Stop

- This STOP button acts as an emergency weld cycle stop.
- If STOP button glows, it indicates that the Arc Welding Robot reaches.
- Its maximum movement in that particular direction.
- Once the STOP glows, the joystick movement will be seized unless
- Reset the STOP button AGAIN.

9.8. Example Program

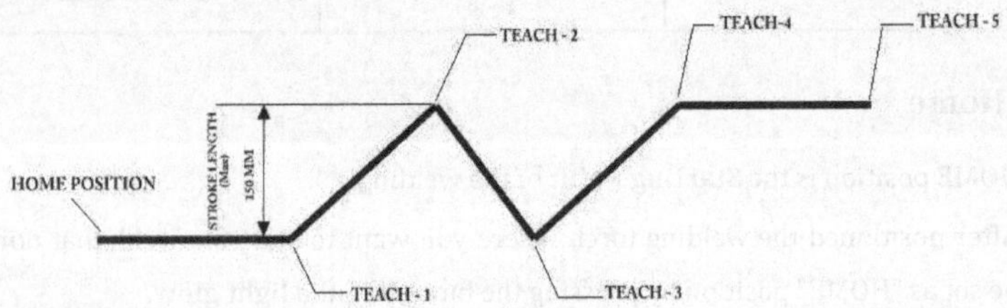

9.9. Procedure

Step 1

Switch ON the PLC control system and also Welding Power Source.

Step 2

You have to move JOYSTICK-1 & JOYSTICK-2 in the appropriate direction to reach the exact position where the welding is to be initiated.

Step 3

After reached the exact position, Press the HOME button till it glows. This point is the HOME position for the welding operation.

Step 4

From the HOME position, you can move the welding robot arbitrarily to reach the "TEACH-1" by using JOYSTICK-1 and JOYSTICK-2. The TEACH-1 is the first reference point for the robot to get along with the weld path. To select "TEACH-1" as a reference point, Press the TEACH button till the light off for fewer seconds.

Step 5

From TEACH-1 point, you can move the welding robot arbitrarily to reach the "TEACH- 2" by using JOYSTICK-1 and JOYSTICK-2. To select "TEACH-2" as a second reference point, Press the TEACH button till the light off for fewer seconds.

Step 6

Repeat the STEP 5 for the remaining TEACH points, say TEACH-3, TEACH-4 and TEACH-5. For the "final TEACH" i.e TEACH-5 , long press is required on TEACH button till the home glow.

Step 7

Press the HOME button to bring back the welding robot to its HOME position.

Step 8

Once the welding torch arrive the HOME position, the start button will glow. This is an indication that the robot is READY to perform the welding in the pre-defined path. Press the start button to perform the welding.

Step 9

At end of the welding, welding robot will be remained at the last Teach point, i.e. TEACH-5. Press the HOME button to bring the torch welding robot to its HOME position

Step10

Repeat the Step 9 for the deposition of additional layer to complete the welding.

9.10. Flow Chart

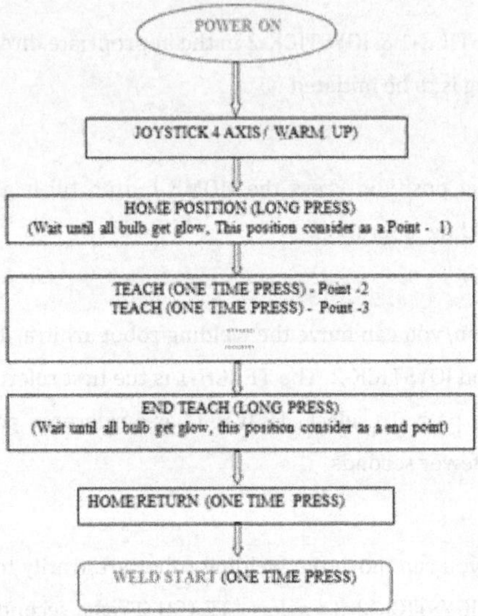

References

1. Fei Gao, Qinglan Chen and Lanzhong Guo, "Study on Arc Welding Robot Weld Seam Touch Sensing Location Method for Structural Parts of Hull", International conference on control, automation and information sciences, Changshu, China, 2015.
2. Lan Hu, Tao Zuwei and Jian Xiaoxia, "Robot welding technology on typical joints of construction machinery", Research and Exploration in Laboratory, Shanghai, Vol. 31, Pp. 15-18, 2012.
3. Hou Jianwei and Wu Yapeng, "Application of medium and heavy plate arc weld robot in rear axle box", MW Metal Forming (semimonthly), Beijing, 2011.
4. Cao Lingyuan, Zhao Bojian and Lin Tao, "Development of welding robot system for ship pipe", Marine Technology, 2008.
5. Yin Fuyuan, Wang Jianli and Jia Zhaoqiang, "Modern automatic welding system for rotating platform robot", MW Metal Forming (semimonthly), Beijing, 2010.
6. A. Kumar and T. Deb Roy, "Heat Transfer and Fluid Flow during Gas-Metal- Arc Fillet Welding for Various Joint Configurations and Welding Positions", Metallurgical and Materials Transactions, Vol. 38, No. 3, Pp.506-519, 2007.
7. Ligong Sun, Fei Xiang, Xiangwen Sun and Sujuan Li, "Design of Industrial Robot Controller Based on System on Programmable Chip", International Conference on Electronic & Mechanical Engineering and Information Technology, 2011.
8. Kailiang Zhanga, Zize Lianga, En Lia and Degang Yang, "Research on an arc welding robot controller with open architecture and the swing-welding interpolation algorithm", International Conference on Advanced Technology of Design and Manufacture, 2010.